Rendezvous in Space

THE SCIENCE OF COMETS

Rendezvous in Space
THE SCIENCE OF COMETS

John C. Brandt
Laboratory for Atmospheric and Space Physics
University of Colorado

Robert D. Chapman
Allied-Signal Aerospace Company
Bendix Field Engineering Corporation

W. H. Freeman and Company
New York

Library of Congress Cataloging-in-Publication Data

Brandt, John C.
    Rendezvous in space : the science of comets / John C. Brandt,
Robert D. Chapman.
      p. cm.
    Includes bibliographical references and index.
    ISBN 0-7167-2175-9
    1. Comets. I. Chapman, Robert Dewitt, 1937–   II. Title.
QB721.B82   1992
523.6—dc20   91-41170
                                    CIP

Printed in the United States of America

1 2 3 4 5 6 7 8 9 0 VB 9 9 8 7 6 5 4 3 2

Contents

Preface

Since we last sat down to write a summary of what is known about comets, there has been a revolution in the field, brought about by the armada of spacecraft that flew to the vicinity of Halley's comet in 1986. The incredible wealth of data collected by the spacecraft that rendezvoused with Halley's comet and comet Giacobini-Zinner validated theories that cometary reseachers had developed over the years. Of course, some new and interesting puzzles were uncovered as well. The long-term careers of comet scientists are quite safe: there remains much to learn.

We have tried to achieve two major objectives in writing this book. First and foremost, we have attempted to summarize the state of knowledge in cometary science as of the beginning of 1992, in a narrative that is accessible to the nonscientist. When describing cometary research, however, one finds phenomena that, to be fully explained, require some understanding of physics, chemistry, and mathematics. Furthermore, the instruments on board the spacecraft that flew through Halley's comet and comet Giacobini-Zinner were very sophisticated. To understand the data from these instruments requires a background in physics and chemistry. Yet, our second objective has been to present as complete a picture of comets as possible. Our struggle between these two objectives led to a compromise: we have introduced enough physics and chemistry in the text to make the narrative understandable, and we have put the mathematics and the data that went beyond a certain level of technical difficulty into two appendices. Armed with the material in these appendices and our annotated bibliography, readers who wish to do so can dig more deeply into the subject. We hope that those who do not want to dig that deeply will not be frightened away. The basic text is an excellent introduction to comets and can be enjoyed with or without the more in-depth material.

Chapter 1 provides a broad description of comets, then discusses the history of our understanding through the Renaissance. Chapter 2 brings the story up to 1950, a watershed year in the field: Fred Whipple, Jan Oort, and Ludwig Biermann put forward theories that revolutionized our understanding of comets. These theories, and the impact they have had on cometary science, are described in Chapters 3, 4, 7, and 8. In Chapter 5 we talk about the modern techniques used to observe comets, including space observations. Chapter 6 focuses on new and exciting material—what we learned when we actually sent instruments into deep space to study two comets. Chapter 9 describes one possible set of future scenarios in comet science; the future will undoubtedly be affected by budget crises, both in the United States and worldwide.

For those who might like to calculate the motions of comets, we have included computer programs to do this in an appendix. The annotated Pascal program was developed using Turbo Pascal version 5.5 and version 6.0. Although it has been tested fairly thoroughly, we cannot guarantee that it is bug-free. The BASIC program is an outgrowth of an Applesoft BASIC version. It now works in GWBASIC, but no effort has been expended to optimize it.

Both of us continue to be excited about comets and hope we have been able to communicate some of that excitement to the reader. We would like to thank Dr. Donald Yeomans for a thorough technical review of the text. He caught a number of our slips; any that remain, however, are ours and ours alone.

John C. Brandt
Estes Park, Colorado

Robert D. Chapman
Columbia, Maryland

January 1992

Rendezvous in Space
THE SCIENCE OF COMETS

1

Introduction: A First Look at Comets

We humans have always been fascinated by comets. The appearance of an especially bright comet excites our interest and stimulates research on such phenomena. When the most famous comet of them all, Halley's comet, came round again in 1985 and 1986, public interest was intense, and the scientific community sent six spacecraft—"Halley's Armada"—to probe its secrets.

Comets typically exhibit a complex variety of physical processes over time and distance. They are superb plasma physics laboratories and are natural probes of the solar-system environment through which they move. It is likely that comets represent material left over from the origin of the solar system; they are "Rosetta stones" that help researchers decode the processes that led to the birth of the sun and its retinue of planets. Comets may even contain complex organic molecules that are fundamental in the chemical processes that lead to life.

What Comets Look Like

Many of us remember reading and hearing about Halley's comet when we were children. This famous comet was a spectacular sight in the night sky when it appeared in 1835 and again in 1910 (Figure 1.1). You may remember the disappointment when the experts who use computers to predict the future motions of comets announced that when Halley passed near the earth and sun in 1986, its path would not be favorable for viewing from the Northern Hemisphere. The calculations showed that when the comet was closest to the sun and at its brightest, the earth would be on the other side of the sun, so Halley would be hidden in the sun's glare. The comet's motion was predicted to bring it relatively close to the earth twice: as it approached the sun (November 27, 1985) and as it began to recede (April 11, 1986). Viewers in the Northern Hemisphere would get their best view in the early days of 1986, after the comet passed close to earth on its approach to the sun. The comet would be bright again in March and best observed from the southernmost United States or the Southern Hemisphere before it passed close to earth on its journey away from the sun.

Figure 1.1 Halley's comet on May 13, 1910. The tail stretches approximately 45 degrees across the sky. At lower left the lights of Flagstaff, Arizona, appear as streaks because the telescope has been moved to match the movement of the comet. Venus is the bright object between the streaks and the comet. (Lowell Observatory photograph)

The comet was predicted to be difficult to see in the large cities of the northeastern United States, where a large portion of the country's population lives, because of the combined effects of its southern location and the glow of city lights. Near Houston, Texas, however, we were able to observe the comet just at the end of twilight for a period of about ten days in January 1986, while it was situated in the constellation Aquarius. It looked like a fuzzy star to the naked eye, but with a pair of 7×50 binoculars it was clearly discernible as a comet. The comet's head was nearest the horizon, not far from the point where the sun had just set. Its tail stretched upward toward the zenith, the point in the sky directly overhead. The whole comet was quite faint, but it was obvious that the head was the brightest part, and the brightness of the tail appeared to decrease as the viewer looked farther from the horizon. Photographs of the comet taken with a 35mm single-lens-reflex camera verified these visual impressions.

We had excellent views of Halley's comet in early April 1986 from Puerto Rico, where the comet and about 5 degrees of its tail—equivalent to 10 lunar diameters—were easily visible to the naked eye (see Plates 1 and 2).

One of the many objectives of this book is to help readers find the information they need to keep track of comet news. Ten to 20 comets are discovered every year, and roughly 20 are observable at any given time. Some of them are so faint that they can be seen only through a large telescope, but hardly a year goes by without at least one comet becoming bright enough to be seen with the naked eye. Some displays have been spectacular. Yet most people are unaware of a bright comet over their heads unless it receives special media attention.

The typical comet has four parts. The brightest part of a comet is its *head*. The *coma* is the head exclusive of the apparent nucleus, a starlike point of light sometimes visible in the head; the solid *nucleus* is discussed on page 5. The comet's *tail* extends away from its head, always pointing away from the sun. Actually, a comet has two tails, one made up of dust and the other composed of electrified material called plasma, each with its own physical processes at work. Comets often display tails of both kinds simultaneously. (See Figures 1.2 and 1.3.)

Comets are best observed through a small wide-angle telescope. Such telescopes are inexpensive; with a little care and

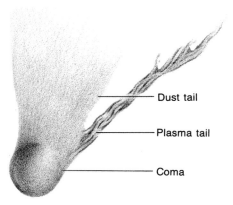

Dust tail

Plasma tail

Coma

Figure 1.2 Compare this drawing of a comet with the photograph shown in Figure 1.3.

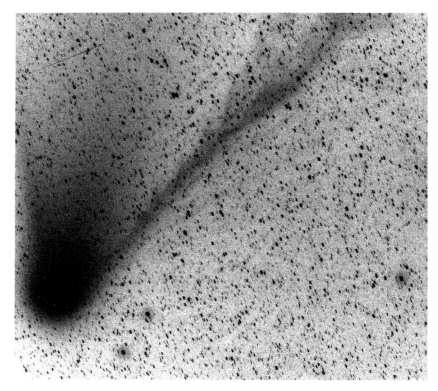

Figure 1.3 Halley's comet on April 18, 1986. Compare this photo with Figure 1.2. (Photograph by William Liller, Island Network, Large-Scale Phenomena Network, International Halley Watch)

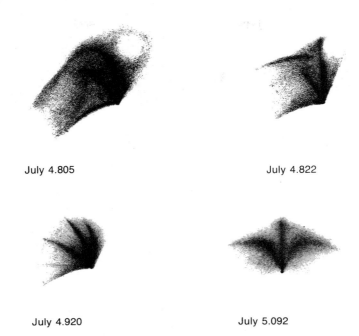

July 4.805   July 4.822

July 4.920   July 5.092

Figure 1.4 Drawings of comet Tebbutt in 1861, showing spiral structure. (From J. Rahe, B. Donn, and K. Wurm, *Atlas of Cometary Forms* [NASA SP-198, 1969])

patience they can even be made at home. People who observe comets through a wide-angle telescope find them to have a fascinating range of internal structures that are in constant motion. Spiral and fountainlike shapes (see Figure 1.4), which seem to originate in the nucleus, can change from hour to hour. With proper care, these structures can be photographed. In a typical comet photograph the exposure is optimized to show the faint tail; the result is a gross overexposure of the much brighter coma. When a photograph is optimized for the coma (as in Figure 3.10, for instance), the tail appears to be truncated.

What Comets Are

The solid part of a comet, the *nucleus,* is a giant snowball that travels around the sun in an orbit that usually differs from a

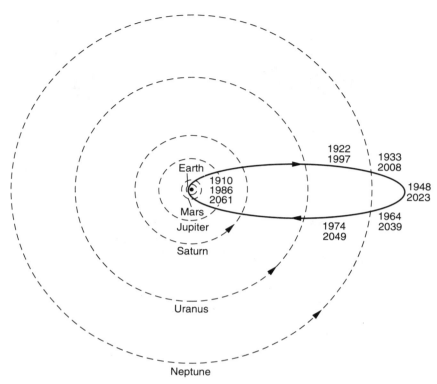

Figure 1.5 The orbit of Halley's comet. The comet travels around the sun in the direction opposite that of the planets and in an orbit that is much more elongated.

planet's orbit; the most obvious difference is that comets' orbits are much more elongated (Figure 1.5). (We will have a lot more to say about this and other differences later.) The nucleus consists primarily of water ice, although the water is not pure H_2O. It contains small quantities of frozen ammonia, carbon dioxide, and a variety of more exotic compounds. Embedded in the ice are solid particles that are thought to be the origin of meteors. This concept of the nucleus, first proposed by Fred Whipple in the 1950s, is known as the dirty-snowball model.

The snowball nucleus spends most of its time in the portion of its orbit farthest from the sun. When it moves close to the sun, the ice sublimates—that is, it passes directly from the solid state to the gaseous state. The released gas, carrying with

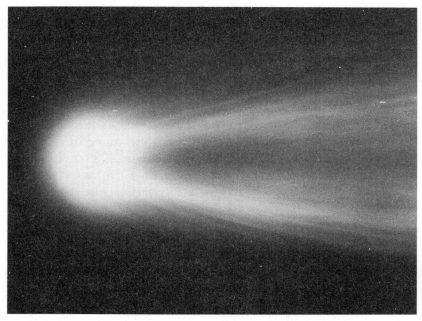

Figure 1.6 The head region or coma of Halley's comet on May 8, 1910, at left. The tail flows away to the right. (Mount Wilson Observatory, Carnegie Institution of Washington)

it some of the solid particles that were embedded in the ice, forms the coma (Figure 1.6). As the comet approaches the sun, the sublimation increases and the coma grows larger, reaching a diameter of 100,000 kilometers or more. The density of material in the coma is always greatest near the nucleus; the farther from the nucleus, the less dense the coma. When we see a starlike point of light in the head of a comet through a telescope, we may be seeing the densest, brightest part of the coma rather than the nucleus.

How Comets Are Discovered

To understand how comets are discovered, we first need to know something about their motion through space. In 1609 Johannes Kepler, in *Commentaries on the Motion of Mars,* stated the first two of his three famous laws of planetary motion. Briefly, he said that (1) the planets move around the sun in elliptical orbits with the sun at one focus; and (2) a straight line between the moving planet and the sun sweeps over equal areas in equal intervals of time. The second law is called the law of areas. (Kepler's laws are discussed in more detail in Chapter 2.)

These two laws apply to any body—such as the earth or a comet—moving under the gravitational influence of a second body, such as the sun. The law of areas describes how the speed of an object orbiting the sun changes as it traverses its orbit. The object moves slowly when it is at *aphelion*, the point in its orbit farthest from the sun, and very rapidly when it is at *perihelion*, the point closest to the sun.

Comets are often discovered as astronomers look for minor planets or *asteroids*—small bodies that orbit the sun mostly between the orbits of Mars and Jupiter—by photographing the region of the sky opposite the sun, called the *opposition region*. Minor planets are relatively easy to see in this region because they are closest to earth and relatively bright. A typical long-exposure, wide-angle photograph of the opposition region of the sky might show the faint trails of a dozen or more minor planets (Figure 1.7). The camera is moved to track the stars closely and record the star images as points. Because the minor planets move along their orbits, they are recorded as streaks or trails.

In 1973 the Czech astronomer Luboš Kohoutek, at the Hamburg Observatory, was making wide-angle photographs of the opposition region to search for "lost" minor planets. Over the course of two weeks he found two previously unknown, faint, fuzzy images on his photographic plates. Kohoutek suspected that the fuzzy images were comets because that is what comets look like when they are far from the sun. The next night he took another photograph of the same region of the sky. If the fuzzy images were once again visible but had moved from their positions of the previous night, then it was very likely that they were comets. That is exactly what he did find. Kohoutek announced his discoveries in telegrams to the Central Bureau for Astronomical Telegrams in Cambridge, Massachusetts. We will see in Chapter 5 that the second of the two comets caused a bit of a stir.

Figure 1.7 A star field with the trailed image (or streak) of an asteroid. (Yerkes Observatory)

The problem with searching for comets far from the sun is that they are likely to be quite faint. The observed brightness of such an object is determined by the amount of energy it reflects that is intercepted by a unit of area (such as a square centimeter) of the observer's eye or of a photographic film in a given unit of time (say, a second).

Imagine several concentric spheres drawn around a comet and assume that no light is absorbed by dust or gas between the observer and the comet. The total amount of energy that passes through the surface of each sphere must be constant. But the observed brightness of the object is not the same at all of the distances to the surfaces of the various spheres. The total energy reflected by the comet is spread out over a larger area of the larger spheres than of the smaller ones, so the energy per square centimeter decreases. Since the surface area of a sphere is proportional to the square of its radius, the amount of energy flowing per second through each square centimeter of the surface of one of the imaginary spheres, and therefore the observed brightness of the source at the surface, must decrease as the inverse square of the distance. This is the simple origin of the inverse square law of brightness, which says that the brightness of an object decreases with the inverse square of its distance from the observer.

The apparent brightness of a comet—or of any other body in the solar system—depends on two distances: its distance from the sun and its distance from the earth. The first distance determines how much light reaches the body and the second how much of its reflected light reaches us. The observed total brightness of a solid body such as Mars depends on the inverse square of each of these distances and, of course, on the reflective properties of the planet. Because a comet expands as it approaches the sun, its brightness typically increases more rapidly than the inverse square of its distance from the sun. Not only does it receive more light as it gets closer, but there is more of it to reflect the light back to the observer. The result is that many—if not most—comets are discovered when they are near both the sun and the earth and are moving rapidly. This is when they are at their brightest and therefore are easiest to spot.

Many amateur astronomers search for comets by sweeping the sky near the sun each evening just after sunset and each morning just before sunrise. A remarkably successful comet hunter who used this technique was the Japanese amateur astronomer Minoru Honda (1913–1990). Honda had been sweeping the skies for comets beginning in 1937 and discovered over a dozen. His reward was to have the comets named after him. Of course, with a dozen comets named after Honda, or two comets named after Kohoutek in a two-week period, we must have specific rules to avoid confusion.

How Comets Are Named

Three distinct schemes for naming comets are in use today. The first is to name a comet after the person or persons who discovered it. If several observers find a comet at more or less the same time, up to three names can be assigned. In 1948, for instance, Honda codiscovered a comet with Antonin Mrkos and Ludmilla Pajdušáková of Czechoslovakia; it was given the name Honda-Mrkos-Pajdušáková. The second scheme is to add a letter designating the order in which the comet is discovered after the year in which it is discovered. Honda-Mrkos-Pajdušáková, for instance, happened to be the 14th comet discovered in 1948, so it was also called comet 1948n, n being the 14th letter of the alphabet. The two comets that Kohoutek discovered were called 1973e and 1973f in this scheme. If more than 26 comets are discovered in a single year, the designation scheme continues with a_1, b_1, c_1, and so on. If more than 52 comets are discovered, the scheme continues with a_2, b_2, and so forth.

The third designation scheme is based on the order in which the comets pass perihelion, the point closest to the sun in their orbit. The first comet to pass perihelion in a given year is assigned the roman numeral I, the second II, and so on. In this scheme, comet 1948a would ordinarily also be referred to as 1948 I. But sometimes a comet does not pass perihelion in the same year it is discovered. The first comet to pass perihelion in 1948, comet 1948 I, was discovered in 1947 as comet 1948k; comet 1948a became 1948 II. The second comet discovered in 1948, comet 1948b, was discovered after it had passed perihelion; it was called comet 1947 XIII. After a comet passes perihelion and takes up a roman numeral designation, the letter designation is seldom used.

A periodic comet, one that has been observed in the inner solar system two or more times, carries the designation P. Thus Halley's comet at the last apparition could be called P/Halley, 1982i, or 1986 III.

Early Knowledge of Comets

A review of the knowledge of comets from the earliest times to the mid-16th century reveals ideas that may seen ludicrous from our vantage point but that were taken seriously in their day and constituted important steps in the process of understanding comets. As we will see, the evolution of cometary thought is intertwined with progress in astronomy as a whole.

The Ancient View of Comets

We must mix a little imagination with a large measure of knowledge of human nature and archaeoastronomy (the study of astronomy in civilizations that existed before written records) to arrive at a picture of what astronomical knowledge might have been like in prehistory. Around 20,000 B.C. the last of the great glaciers were receding and the giant mammals (the saber-toothed cat and the mastodon of North America and the mammoths and other giants of Europe and Asia) were slipping into natural extinction. Our ancestors produced drawings of animals of breathtaking beauty in the caves of central and western Europe. These people depicted what was important to them—the animals that provided the meat, skins, bone, and sinew that they used for sustenance, clothing, and the implements of everyday living.

They left us no likenesses of anything that we would call astronomical, yet these people must have had a knowledge of the sky. They certainly were aware that the annual changing of the patterns of stars—or constellations—signaled the changing of the seasons. They probably were able to tell time and the seasons by the position of the sun. The changes in the night sky—the daily risings and settings and the annual progressions—are orderly processes with which they could hardly fail to become familiar. We can only guess their reaction to an object in the sky that did not fit into the natural order, such as an eclipse of the sun or moon or the appearance of a bright comet. If one of these celestial events coincided with a terrestrial event that was particularly noteworthy—an especially successful hunt or the death of a leader, say—they might well have ascribed a cause-and-effect relationship to the events.

The more recent cultures that evolved as humankind shifted from a hunting to an agrarian way of life had a surprisingly well-developed knowledge of the sky. The first records we have in many areas are quite advanced. Among the ancient civilizations, the Egyptians, the Babylonians, and the Chinese all knew the numerical values involved in the motions of the sun, moon, and planets with sufficient accuracy to predict their positions. This knowledge of motions and predictions probably developed from their need to determine their calendar. In any event, the tables of the motion of Ishtar (Venus) compiled by the Chaldeans exhibit detailed familiarity with the sky. Yet we can find few references to comets in the writings that have survived from those times.

The earliest writings in which we find detailed discussions of comets date from the time of the great Greek philosophers. Plato (428–347 B.C.) had a well-developed concept of the nature of the universe. We see little of it in his writings, but he stimulated his students to develop their own views, their own cosmologies. One of his first students was Eudoxus of Cnidus. Eudoxus imagined the universe to be made up of 33 spheres, all centered on the earth and interconnected in such a way that the rotation axis of any sphere was attached to the surface of the next larger sphere. Each sphere in the complex rotated at a certain speed. By choosing the orientations and speeds of the spheres appropriately, and by attaching each of the five planets that can be seen with the naked eye to the correct sphere, Eudoxus built up a theoretical model that mimicked the motions of the planets very well. Eudoxus did not propose that the spheres represented reality. For him the model was merely a convenient mechanism—like our modern mathematical description of planetary motions—for predicting the future positions of the planets.

Perhaps the most famous student at Plato's Academy was Aristotle (384–322 B.C.), who came to study with Plato in 367 B.C. and remained for 20 years. We know a great deal about Aristotle's life, including the fact that he was hired by Philip of Macedonia to tutor Philip's young son—the lad who grew up to be Alexander the Great. Aristotle wrote extensively on mathematics, logic, astronomy, biology, and many other topics. He expanded on Eudoxus's conceptual model of the planetary system in two ways: he added 22 more spheres, and he strongly implied that the spheres were real and that they caused the planets' movement.

Aristotle described his views on the nature of comets in his book on meteorology. He assigned comets (and also the Milky Way) to the upper regions of the earth's atmosphere. Aristotle believed that all matter is composed of four substances or elements: fire, air, water, and earth (Figure 1.8). According to him, each of these basic elements could originate from the others and each contains all the others. Incidentally, later writers introduced a fifth substance or *quinta essentia,* the pure substance that makes up the bodies in the heavens; hence our word *quintessence.*

According to Aristotle, when the sun warms the earth, the earth's substance evaporates into two different kinds of vapors: a moist, cool vapor and a warm, dry, windy vapor. The windy exhalation rises above the moist vapor until it reaches the edge of space, which Aristotle called "the outermost part of the terrestrial world which falls below the circular motion." The circular revolution then carries the warm exhalation around the earth. "In the course of this motion it often ignites wherever it may happen to be of the right consistency, and this we maintain to be the cause of [shooting stars]." A comet is formed when the circular motion introduces into the warm exhalation just the proper level of fire—enough to be a shooting star but not enough to ignite everything. "The kind of comet varies according to the shape which the exhalation happens to take. If it is diffused equally on every side the star is said to be fringed; if it stretches out in one direction it is called bearded." Aristotle went on to argue that whenever a comet appears in the sky, it foretells wind and drought. This fact, he said, clearly indicates the "fiery constitution" of comets and shows that they arise from drier air.[1]

Many other philosophers wrote about the nature of comets in the centuries after Aristotle, and we can see his influence in some of their writings. Heraclides of Pontus, who was a student of both Plato and Aristotle, wrote that a comet was an

---

1. Aristotle gives a detailed description of comets in his *Meteorology*. This and the preceding quotes are from that work. The translation we used is published in *Great Books of the Western World*, vol. 8, pp. 445–494 (Encyclopedia Britannica, 1952).

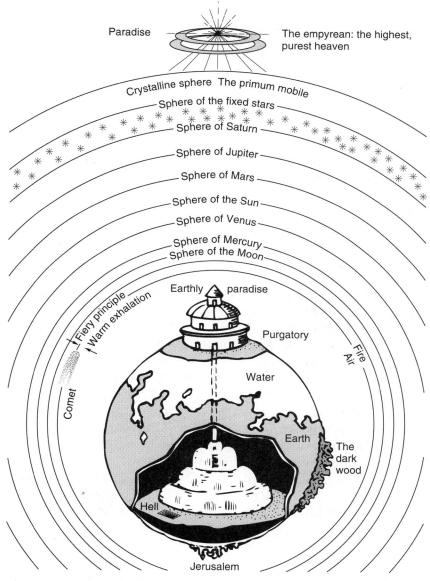

Paradise

The empyrean: the highest, purest heaven

Crystalline sphere The primum mobile

Sphere of the fixed stars

Sphere of Saturn

Sphere of Jupiter

Sphere of Mars

Sphere of the Sun

Sphere of Venus

Sphere of Mercury

Sphere of the Moon

Fiery principle

Warm exhalation

Earthly paradise

Purgatory

Comet

Fire

Air

Water

Earth

The dark wood

Hell

Jerusalem

Figure 1.8 The Aristotelian view of the universe as "modernized" by Dante in *The Divine Comedy*. We have simplified the figure and added the Aristotelian theory of the origin of comets. Comets (left) were supposedly formed when fire interacted with a warm exhalation from the earth. (After R. D. Chapman and J. C. Brandt, *The Comet Book* [Boston: Jones & Bartlett, 1984])

inflamed cloud high in the atmosphere. Other thinkers believed that comets were celestial bodies. Apollonius of Myndus, for example, believed that a comet is a distinct type of celestial body that becomes visible only when it "reaches the lowest portion of its course."[2] Apollonius's discussion of comets has some gems of truth in it.

We should not conclude, however, that Aristotle was wrong and Apollonius was right. In fact, the only observational evidence that either of them had to work with was the visual appearance of numerous comets, and their theories were largely guesswork. As we shall see, it was not until the 16th century that firm scientific evidence on the celestial nature of comets was obtained.

Comets in Medieval Times

The Venerable Bede (673–735) was one of the more important writers of medieval Europe. He grew up in England in a Benedictine monastery in Northumbria (now Northumberland). Fortunately for him, the monastery had an outstanding library, and he read widely as he matured. Bede wrote nearly three dozen books, two of which touch on comets. In his *De Rerum Natura (On the Nature of Things)* Bede described comets as stars with fiery hair that were portents of great happenings—war, pestilence, the deaths of kings, and other dread events. He did assert that at least some comets move through the celestial regions like planets.

Bede also talked about comets in his *Historia Ecclesiastica*, which historians consider to be one of the best sources of medieval English history up to the time of his death. Two bright comets appeared in January 729 and remained visible for about two weeks, one in the evening sky and one in the morning sky. The two comets on opposite sides of the sky, Bede said, "portended awful calamity to east and west alike. Or else, since one comet was the precursor of day and the other of night, they

---

2. Apollonius's description of comets is quoted in Seneca, *Questions Naturales*, trans. J. Clarke (London: Macmillan, 1910).

indicated that mankind was menaced by evils at both times."[3] Bede blamed the death of King Osric of Northumbria on the appearance of the comets. (Incidentally, the monasteries of Northumbria were destroyed by the Vikings a few years after Bede's death. Bede was canonized as Saint Bede the Venerable in the 19th century.)

Many other thinkers of the Middle Ages and early Renaissance, including Thomas Aquinas and Roger Bacon, also wrote about comets. We will not discuss their work here, but we cannot leave this period without mentioning two beautiful works of art recording the comet that would become known as Halley's comet. The first is the famous Bayeux tapestry (Plate 3), which was probably created between A.D. 1070 and 1080. It depicts the Norman conquest of England in 1066, when William the Conqueror defeated King Harold. One panel of the tapestry shows the comet, Harold and his adviser, and some of his subjects. One can imagine that Harold's adviser is telling him that the comet portends bad news. (On the other side of the English Channel, William's advisers may have been telling him that the comet indicated good news.)

The second work of art is by Giotto di Bondone (ca. 1276– 1337), who is often regarded as the first artist of the Italian Renaissance. His life certainly spanned the period when Italy was moving out of the Middle Ages into the full flower of the Renaissance. Among Giotto's masterpieces are frescoes he painted on the walls of a small chapel in Padua, built by the Scrovegni family (Plate 4). The story goes that Enrico Scrovegni built this chapel and hired Giotto to decorate it to atone for his father's wicked life. (The father, Reginaldo Scrovegni, was apparently infamous enough to be the usurer referred to in Canto XVII of the *Inferno,* where Dante speaks of "one who had his little white sack marked with an azure and gravid sow." That azure and gravid sow appears in the coat of arms of the Scrovegni family.) One of the frescoes, *The Adoration of the Magi,* shows the Three Wise Men presenting their gifts to the Christ Child. Above the manger the star of Bethlehem looks surpris-

---

3. Bede, *Historia Eccelsiastica,* or *A History of the English Church and People,* trans. L. Sherley-Price (New York: Penguin Books, 1968).

ingly like a comet. The fresco was painted in 1303 and Halley's comet was a bright object in European skies in 1301, so it is likely that Giotto painted Halley's comet as the Christmas star. If this conjecture is correct, it is one of the most magnificent of all the records of the comet.[4]

The Renaissance View of the Universe

Nicolaus Copernicus (1473–1543) is one of the seminal figures in the renaissance of astronomy in Europe. He is important because he set the stage for the work of 16th-century thinkers. Copernicus was a student at the University of Krakow in Poland, where he received an education based strongly on mathematics and astronomy. In *On the Revolutions of the Heavenly Spheres* Copernicus broke with Aristotle and said that the sun is the center of the universe and the earth is merely one of the planets that revolve around it. He believed that the daily rising and setting of the stars and other heavenly bodies are merely a reflection of the earth's rotation on its axis, and the annual motion of the sun among the stars is due to the earth's revolution around the sun. The work of Copernicus caused a furor among the scholars of his time, to whom belief in Aristotle's truth was a matter of faith.

Aesthetics aside, Copernicus had little more evidence for his heliocentric view of the universe than Aristotle and his contemporaries had for their geocentric view. Yet the writings of Copernicus laid the foundation for our modern view of the world. We find some of the greatest thinkers of the 15th and 16th centuries taking up the heliocentric cause—such men as Francis Bacon, René Descartes, and Thomas Digges. There were some setbacks. In 1600, for example, Giordano Bruno was burned at the stake for espousing the Copernican philosophy, and such pragmatic men as Descartes decided that it was prudent to keep quiet on the issue. Copernicus's view was gathering momentum, however, as observational evidence supporting it mounted.

---

4. Art historian Roberta Olson discusses comets in art more broadly in an interesting article, "Giotto's Portrait of Halley's Comet," *Scientific American* 240 (May 1979):160–170. Olson has also written *Fire and Ice, A History of Comets in Art* (Washington, D.C.: National Air and Space Museum, 1985).

2

The Orbits and Motions
of Comets

Development of Modern Ideas about Comets

Until the Renaissance, our ideas about the planets and other celestial bodies were based on visual observations and some measurements made with relatively crude devices such as the cross-staff. Developments in the 16th century improved on this situation and began the process that would lead to our modern ideas about the universe.

Among the advances that went hand in hand with the development of modern astronomy was the development of the tools that astronomers needed to do their complex computations. New, more precise tables of trigonometric functions were calculated and new trigonometric formulas were devised. The invention of logarithms and the calculation and publication of tables of logarithms at the beginning of the 17th century greatly expedited the calculation process.

Tycho Brahe and the Comet of 1577

The Danish nobleman Tycho Brahe made significant improvements in the instruments used to measure the positions of objects in the sky. Special care in the methods of construction and attention to many details, such as the markings used to read the positions, led to the development of sextants (used to measure the distance between two stars) and quadrants (used to measure the altitude of stars).

Tycho was born in 1546 and was educated at the University of Copenhagen and the University of Leipzig, which were among the foremost seats of learning in Europe. His uncle died in 1565 and left the young Tycho quite affluent. With his own funds and an endowment from King Frederick II of Denmark, Tycho set up an observatory, Uraniborg, on the island of Hven near Copenhagen. Here Tycho fashioned an impressive array of instruments to make accurate measurements of the sky. All this activity began more than 25 years before the telescope was invented, and Tycho made all his measurements without the benefit of any optical devices.

Before we can understand the relevance of Tycho's measurements to comets, we have to understand the concept of parallax (Figure 2.1). If you look at a nearby object against the background of more distance objects and then change your position, you will see the nearby object shift against the background. This is something we make use of all the time, more or less consciously, to gauge distances. If you move far enough from your original position, even celestial objects appear to shift. This is the effect known as *parallax*. If you observe the moon from two locations separated by thousands of miles, the moon's position in relation to the more distant stars can be seen to shift. To make this observation, it is not necessary to move the measuring instrument around the globe physically; just wait several hours and the rotation of the earth will do the moving for you.

In 1572 a new star appeared in the constellation Cassiopeia. The *nova* (which was actually what we call a supernova today, the incredibly violent "death rattle" of a star) became the brightest object in the sky for a short time—so bright that it was visible even in broad daylight. Tycho measured the position of the nova in relation to the other stars in Cassiopeia

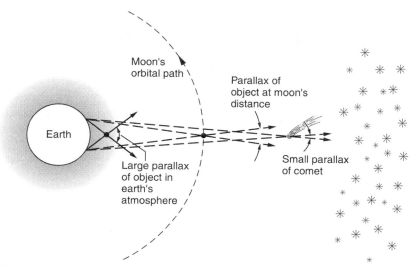

Figure 2.1 This diagram illustrating the concept of geocentric parallax (not drawn to scale) shows the parallax angle at three distances from the earth. Notice that as the distance from the earth increases, the parallax angle decreases; the farther from the earth, the smaller the angle. By measuring this angle for the comet of 1577, Tycho Brahe established that the comet was well beyond the moon. (After R. D. Chapman and J. C. Brandt, *The Comet Book*, [Boston: Jones & Bartlett, 1984])

during the time it was visible, hoping to detect a parallax. When he was unable to detect any shift, he concluded that the object was much farther away than the moon. This was a rather more significant conclusion than it sounds at first, because the Aristotelians believed that the heavens—except for the planets—were immutable. Yet here was a clear observation of a change. The move to our modern view of the universe had begun.

In November 1577 a bright comet appeared in the skies over Europe and remained visible for several months. Tycho and a contemporary, Hagecius (ca. 1526–1600), attempted to measure the parallax of the comet. Tycho's observations showed that the parallax of the comet was about one-fourth as large as that of the moon, so he concluded that the comet was at least four times farther from the earth than the moon is. Hagecius announced a parallax of 6 degrees, six times larger than the moon's, making the comet very near the earth. Tycho was eventually able to show that Hagecius had misinterpreted his own observations, and that they were in fact consistent with his own.

Once Tycho knew the comet's distance, he could calculate its size. He measured the tail as 22 degrees long and 2.5 degrees wide and the head as 8 minutes of arc across. This meant that the tail was 350,000 kilometers long and the head was over 2000 kilometers in diameter. The resolution of the disagreement demonstrated that the comet of 1577 was not an Aristotelian apparition in the earth's atmosphere but a large celestial body. If that was true of one comet, Tycho concluded, then it was probably true of all of them.

Tycho had an abrasive personality and was a poor manager, and eventually he wore out his welcome at the Danish court. By 1597 Frederick II of Denmark had died, and the new king was not sympathetic to Tycho. Tycho therefore offered his services as court astrologer to Emperor Rudolf II and moved to Prague. There he advertised for an assistant to help carry out his observations and to do the calculations required to interpret his voluminous earlier observations. Johannes Kepler (1571–1630) came to Prague and spent years working on the data, first as Tycho's assistant, then alone.

Tycho died in 1601, following a dinner with a group of nobles. He was too polite to leave the table to relieve his bladder. Kepler describes several days of delirium, suggesting that Tycho died of extreme urinary complications.

Kepler and the Laws of Planetary Motion

Johannes Kepler was born near Stuttgart in 1571 and was educated at the University of Tübingen. He then settled as a schoolmaster in Graz, Austria, where he married and had a houseful of children. As a Protestant, however, he was forced from Graz by Archduke Ferdinand. It was at this point that Kepler moved to Prague to become Tycho's assistant. When Tycho died, Kepler was appointed his successor, with a salary from the emperor and access to Tycho's data.

From a detailed study of Tycho's excellent observations, Kepler was able to formulate three fundamental laws of planetary motion. He began by demonstrating that the orbit of Mars was an ellipse, with the sun at one focus. By the process of inductive reasoning—that is, arguing from the particular to the general—Kepler postulated that all planets move in elliptical orbits. He could check this idea by comparing the orbit of Mars with the orbits of other planets. All tests of this hypothesis confirmed it. Kepler also noted that the planets move fastest in their orbits when they are closest to the sun and slowest when they are farthest from it. These conclusions are formalized in his first two laws of planetary motion, which were published in 1609 in his *Commentaries on the Motion of Mars:*

1. All planets move in elliptical orbits with the sun at one focus (Figure 2.2*a*).

2. A straight line joining the planet to the sun sweeps out equal areas in equal intervals of time (Figure 2.2*b*).

These two laws of motion apply equally well to comets. While the ellipses of planetary orbits are nearly circular, however, those of cometary orbits tend to be highly elliptical. In the most extreme cases, cometary orbits can be virtually indistinguishable from a parabola. It is interesting to note that Kepler did not recognize this fact. He believed that comets moved in straight lines with varying speeds. Given Kepler's extensive work on elliptical orbits, this judgment seems curious. If comets moved on straight lines that passed near the earth, though, they would be seen only when they were close to the earth and only once. Thus their sudden appearance and disappearance could be explained.

Kepler also stated a third law of planetary motion, which he discovered a number of years after the first two:

3. The ratio of the squares of the periods of revolution of any two planets is equal to the ratio of the cubes of their average distance from the sun (Figure 2.2*c*).

At the time this law was discovered, it was simply an empirical relationship. Mars, for example, is 1.52 *astronomical units*, or *AU*, from the sun and revolves around the sun in 1.88 years. (An astronomical unit is equal to the mean distance between the earth and the sun—150 million kilometers, or 93 million miles.) Thus the distance cubed $(1.52 \times 1.52 \times 1.52) = 3.54$, and the revolution period squared $(1.88 \times 1.88) = 3.54$. Here earth is the comparison planet; that is, it is the denominator in the

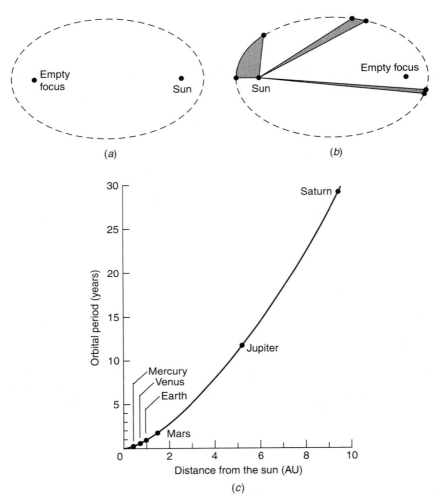

Figure 2.2 Kepler's three laws of planetary motion. *(a)* The planet's orbit is an ellipse with the sun at one focus. *(b)* The straight line from the planet to the sun sweeps out equal areas in equal times. The areas of all three shaded sections are approximately equal. Thus the planet moves faster when it is near the sun than when it is far away. *(c)* The orbital period (in years) squared is equal to the heliocentric distance (in astronomical units) cubed. The curve shown illustrates this relationship. For a comet, the heliocentric distance is replaced by the *semimajor axis*, or one-half the longest dimension of its elliptical orbit.

ratios mentioned in the third law. With this law, simply determining the period gave the distance from the sun in astronomical units.

Although Kepler's laws were consistent with Copernicus's views, they did not constitute proof of the Copernican system. Proof would have to await the first use of the telescope to observe the cosmos.

Galileo

While Kepler was intent on his studies of Tycho's data in Prague, one of the superstars of science was hard at work in Italy. Galileo Galilei was born in Pisa in 1564 and was educated at the university there. His work contributed greatly to progress in understanding the solar system.

Word reached Pisa in 1609 that a Flemish spectacle maker, Jan Lippershey, had mounted a concave lens at one end of a tube and a convex lens at the other end, and so contrived an optical device that would magnify distant objects. What Lippershey had built was, of course, a simple telescope. Galileo made a telescope for himself and turned it toward the heavens. He saw wonders of the universe that no one had ever seen before. One of his important discoveries was that the planet Venus exhibits phases and changes in size. This discovery was entirely consistent with Copernicus's belief that Venus orbited the sun inside the earth's orbit. This single observation was one of the first irrefutable proofs of Copernicus's view of the universe.

Galileo's telescope also revealed that the Milky Way was made up of a large number of faint stars that could not be seen individually with the naked eye, but that together produced the amorphous glow that gives our galaxy its name. The resolution of the Milky Way into individual stars immediately disproved Aristotle's contention that the Milky Way, somewhat like a comet, was an apparition in earth's atmosphere.

The story of how Galileo ran afoul of the church is a fascinating chapter in the history of science that we will not recount in detail here. Galileo, the true scientist, offered his telescope to the disbelievers. "Here," he would say, "look for yourselves, the evidence is incontrovertible." His detractors would reply, "We have no need to look, we have faith that you are wrong. The evidence you claim can't be there." We should not be surprised that many people in the 17th century were so convinced that they knew the truth that they would not confront the facts. One need only read the newspapers today to see that three and a half centuries of "progress" have done little to change the way the human mind works. In the short run, Galileo was silenced by the Inquisition, and world leadership in science moved from Italy to England. In the long run, Galileo has prevailed.

Galileo's views on comets did not contribute to scientific progress, however. In his book *Assayer*, published in 1623, he argued that a comet might be "an appearance rather than a real object," and he simply discounted Tycho's parallax observations of the comet of 1577. The arguments between Galileo and his contemporaries, which led to the publication of *Assayer*, were prompted by the appearance of three comets in 1618. Writings on these comets by Galileo and some of his contemporaries, while interesting as salvos in a scientific skirmish, generated more heat than light. Yet Galileo's work on mechanics, which laid the groundwork for Newton's development of classical mechanics and the theory of gravitation, indirectly did a great deal to further our understanding of comets.

Newton and Halley

Isaac Newton (1642–1727) stands as one of the most intellectually productive individuals in history. He entered Trinity College at Cambridge in 1661, but before he could complete his studies, the university was closed because of the plague. Newton spent two years at his family home while the plague ran its course. In those two years of leisure he invented differential calculus, devised the theory of colors, and began to develop the quantitative science of classical mechanics.

Using his newly invented mechanics and Kepler's laws of planetary motion, Newton arrived at the law of universal gravitation, which states that every body in the universe attracts every other body with a gravitational force that is proportional to their masses and inversely proportional to the square of the distance between them. The planets, comets, and lesser bodies of our solar system do not fall away from the sun because gravity holds them in place. The essential simplicity of the Copernican solar system, recognized by Kepler, had been quantified by Newton, and this view became universally accepted in scientific circles.

By the late 17th century it was well accepted that comets are celestial bodies that move around the sun in parabolic or nearly parabolic orbits. Highly elliptical ellipses closely resemble parabolas. Thus many comets' orbits are described as "nearly parabolic." Newton devised a method to calculate the size and orientation in space of a parabolic orbit using three observations of a comet's position among the stars. This method was to be used by Edmond Halley (1656–1742; Figure 2.3), a brilliant scientist who made significant contributions in several areas.

One of Halley's greatest contributions has to do with the comet that bears his name. Halley used Newton's method to calculate the characteristics of the orbits of roughly two dozen comets that had been well observed over the previous several centuries. The result was quite remarkable: comets observed in 1456, 1531 (Figure 2.4), 1607, and 1682 all seemed to move in the same orbit. Halley concluded that these four comets were

Figure 2.3 Edmond Halley.

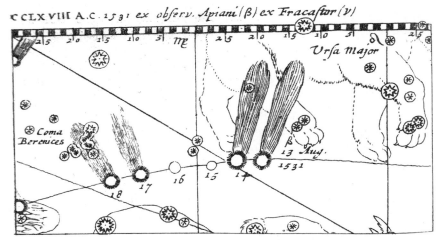

Figure 2.4 Peter Apian's observations of Halley's comet in 1531. The observed orientation of the tail showed that it pointed away from the sun. (*Bulletin de la Société Astronomique de France*, 1910)

in fact the same comet, one that orbits the sun every 76 years. He predicted that it would return again in 1758. Halley was aware that the gravitational attraction of Jupiter and Saturn could affect the motion of the comet and that his prediction did not take these effects into consideration. The French astronomer Alexis-Claude Clairaut (1713–1765) calculated the perturbing influence of Jupiter and Saturn on the comet and predicted that its orbit would pass closest to the sun—the perihelion point—on April 15, 1759. The comet was rediscovered on Christmas night 1758 and passed perihelion on March 13, 1759. Given the uncertainties of calculation at the time, Clairaut had come very close indeed.

To honor Halley's work, the comet was given his name. The success of Halley's prediction did two things: it established the validity of Newtonian mechanics and established the celestial origin and nature of comets beyond any doubt.

Celestial Mechanics

The science of celestial mechanics developed rapidly after Halley's success. Many of the advances made in mathematics by

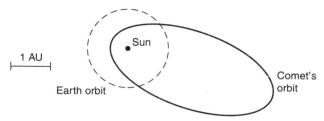

Figure 2.5 The orbit of Encke's comet projected onto the plane of the earth's orbit.

such men as the marquis de Laplace (1749–1827), Karl Friedrich Gauss (1777–1865), and Joseph-Louis Lagrange (1736–1813) were stimulated by the need for very precise calculations if the positions of planets and comets were to be predicted with any accuracy, since their motions are influenced not only by the sun but also by gravitational perturbations of the other planets.

The method that Gauss devised to calculate the orbital characteristics of a body moving in an elliptical orbit was used by Johann Encke (1791–1865) to calculate the orbits of a number of unusual comets discovered in the late 18th and early 19th centuries. No parabolic orbit would fit these observations, but a single comet moving in an elliptical orbit (Figure 2.5) fit them all. Its position varied between 0.34 and 4.08 astronomical units from the sun with a period of only 3.3 years. Thus he came to realize that comets move around the sun in orbits that vary widely in size and shape.

Today we have orbital parameters for well over 1000 appearances of comets. Brian Marsden of the Harvard-Smithsonian Center for Astrophysics periodically publishes the *Catalogue of Cometary Orbits*. His sixth edition, published in 1989, lists 1292 orbits. *Short-period comets,* such as Halley's, are listed numerous times (Halley's, in fact, is listed 30 times). Since planetary perturbations modify the orbital parameters, each listing differs slightly from every other. *Long-period comets,* with periods of more than 200 years, account for 655 entries. The catalogue does not list the periods of most of the long-period comets since their length is uncertain. But many of those periods are certainly millions of years long.

Soon after the short-period elliptical orbit of Encke's comet was verified, another comet, Biela's, was shown to move in an elliptical orbit with a 6.75-year period and an aphelion (far-thest point from the sun) near Jupiter's orbit. The motion of Biela's comet was influenced considerably by Jupiter. Biela's comet is interesting for another reason, however. During its predicted return to the vicinity of the earth and the sun in 1846, the comet split into two distinct comets, which were ob-served to move away from the earth along the same orbit. In 1852 both comets returned on schedule. They then left the vicinity of the earth for what may have been the last time; they have never been seen again.

Occasionally the earth encounters swarms of *meteoroids*— small solid particles that orbit the sun just like tiny planets, all moving in the same path. Scientists can determine the orbits of these bodies by observing the streaks of light they produce when they burn up in the earth's atmosphere. When the earth encounters such a stream, we frequently observe meteor show-ers (Figure 2.6). That is, we see a large number of meteors streaking across the night sky, all seeming to radiate from the same point. By the 1860s it had been established that there was a connection between the orbits of meteor streams and comets. In 1872, when Biela's comet was due to return to earth's vicinity, a spectacular meteor shower was observed radiating from the constellation Andromeda.

By the end of the 19th century, a considerable body of scientific research had been undertaken in an attempt to learn more about the nature and dynamics of comets.

Orbits and Dynamical Statistics

Bodies in the solar system such as planets, asteroids, and com-ets move around the sun in orbits that are *conic sections* (Figure 2.7). The curve or orbit can be a *hyperbola,* a *parabola,* an *ellipse,* or a *circle.* An object starting a one-time pass through the solar system with a speed greater than zero when it is far from the sun will move in a hyperbolic orbit. This type of motion would be characteristic of an interstellar comet. An object starting a one-time pass through the solar system with a speed of zero far from the sun will move in a parabolic orbit. Although the

Figure 2.6 Woodcut of the Leonid meteor shower of 1833. (Courtesy of the American Museum of Natural History)

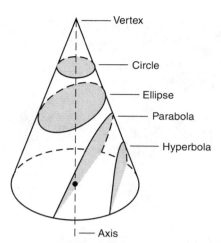

Figure 2.7 The conic sections. The orbit of a comet can take several forms that mathematically depend on the way a plane intersects a cone. A *circle* is produced if the plane is at right angles to the axis of rotation. An *ellipse* is produced if the plane's angle with the axis is between a right angle and the angle necessary to produce a parabola. A *parabola* is produced if the plane makes the same angle with the axis as a line on the surface of the cone passing through the vertex. A *hyperbola* is produced if the planes make an angle with the axis between the angle necessary to produce a parabola and a position parallel to the axis.

orbits of many long-period comets are ellipses, the part of such an orbit that is near the sun is actually very nearly a parabola. The orbit of a newly discovered comet is often assumed to be parabolic for the first approximate calculations. As we have seen, Edmond Halley was one of the first people to use this approximation.

An object on a closed orbit with a changing distance from the sun will move on a path that is an ellipse. A special case of an ellipse, the circle, occurs when the distance from the sun does not change. The degree of elongation of an ellipse is described by a quantity called the *eccentricity*. Greatly elongated, cigar-shaped orbits have eccentricities approaching 1, whereas nearly circular orbits have eccentricities approaching 0 (see Table 2.1). Most cometary orbits in fact are elongated ellipses. The orbit of Halley's comet, for example, has an eccentricity of 0.967, Encke's comet has an eccentricity of 0.847, and comet

Table 2.1 Orbit Shapes

Shape	Eccentricity
Circle	$e = 0$
Ellipse	$0 < e < 1$
Parabola	$e = 1$
Hyperbola	$e > 1$

Schwassmann-Wachmann I has an eccentricity of 0.101. Many planetary orbits, including the earth's, are nearly circles.

To describe the size, shape, and orientation of an elliptical comet orbit in space and a comet's location along this orbit requires the specification of six parameters called the *orbital elements.*

Orbital Elements

All the information required to calculate the future positions of a comet—if we ignore for the moment the perturbations caused by the other planets—is contained in the six orbital elements of the comet (Figure 2.8). Two of the elements describe the size and shape of the orbit, three additional elements describe its orientation in space, and the final element describes a starting position of the comet in its orbit. Celestial mechanicians use more than one set of orbital elements, but the sets are equivalent in the sense that any set can be derived from the others.

The elements used to describe the size and shape of the orbit depend on whether the orbit is parabolic, nearly parabolic, or just highly elliptical. In any case, one of the elements is the eccentricity of the orbit. The eccentricity of a conic section ranges between zero and a large number (Figure 2.9).

In the case of an ellipse, the second orbital element often used is the *semimajor axis* of the ellipse, which is one-half the longest dimension of the ellipse. In the case of a circle, the radius is equivalent to the semimajor axis. For a parabola or hyperbola, the semimajor axis is undefined, since the figures are not closed. In these cases the perihelion distance is used as the second parameter.

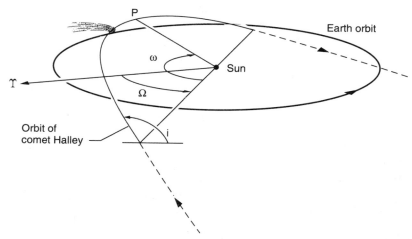

Figure 2.8 The angular orbital elements of Halley's comet. P is the perihelion point, the orbital inclination i is 162 degrees, the longitude of the ascending node Ω is 58 degrees, and the argument of perihelion ω is 112 degrees. The direction of the vernal equinox is indicated by ♈. (After D. K. Yeomans, *The Comet Halley Handbook*, 2d ed. [International Halley Watch, JPL 400-91, 1983])

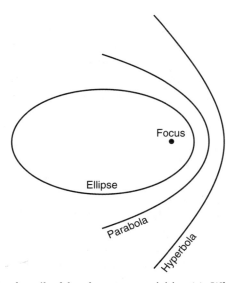

Figure 2.9 Orbits described by three eccentricities *(e)*. When *e* = 0.9, the orbit is the ellipse shown. When *e* = 1.0, the orbit is the parabola shown. When *e* = 1.1, the orbit is the hyperbola shown.

Figure 2.8 shows the three angular elements often used. They are the inclination of the orbital plane to the plane of the *ecliptic*, i (the ecliptic is the apparent path of the sun around the sky); the longitude of the ascending *node*, Ω; and the argument of perihelion, ω. The ascending node is the point on the orbit where the comet passes from below the plane of the ecliptic to above the plane of the ecliptic. These three angles uniquely define the orientation of the orbit in space.

The final element frequently used is the time of perihelion passage; that is, the time at which the comet is closest to the sun. Note that the period of revolution of the comet around the sun is not an orbital element. We can calculate the period of a comet in an elliptical orbit from the orbit's semimajor axis, using Kepler's third law of planetary motion. Given these six elements of the orbit, we can calculate a comet's future positions. In Appendix B we describe a brief program that can be used to do these calculations on a home computer.

The six elements precisely determine a comet's orbit when the orbit is an ellipse. Gravitational perturbations by the planets and the so-called nongravitational forces keep cometary orbits from being exact ellipses. The orbital elements listed in tables are often *osculating elements*—that is, elements of an ellipse that exactly match the comet's position and motion at the instant of osculation (or "kissing"). Around this instant, the osculating ellipse is a good approximation of the comet's orbit.

Orbital Statistics

Patterns in the statistics of the orbits of known comets tell us something about the origins and evolution of comets. Figure 2.10*a*, for example, shows the distribution of the inclinations of the orbits of parabolic and nearly parabolic comets. What this figure tells us is that the inclinations are randomly distributed in space. Compare this figure with Figure 2.10*b*, which shows the distribution of inclinations for short-period comets. Most of the short-period comets move in orbits that lie close to the plane of the planetary orbits. These two statistical distributions lend credibility to Oort's hypothesis, that long-period comets fall out of a spherical cloud of comets; we would expect these comets to come in toward the sun from all direc-

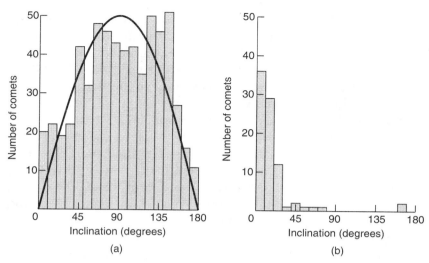

Figure 2.10 *(a)* The distribution of inclinations of the orbits of parabolic and nearly parabolic comets. The curved line indicates the distribution if the inclinations are random. *(b)* The distribution of inclinations for short-period comets. Note that these orbits are direct (in the same direction as the planets) and are mostly near the plane of the earth's orbit. (Data from B. G. Marsden, *Catalogue of Cometary Orbits,* 5th ed. [Cambridge, Mass.: Harvard-Smithsonian Astrophysical Observatory, 1986])

tions. Once a long-period comet is captured in a short-period orbit by the planets, it will move more like the planets.

Figure 2.11 is a histogram of the number of short-period comets in relation to their aphelion distance. Note that there is a spike at 5 astronomical units, Jupiter's distance from the sun. These comets belong to a family of comets whose motion is dominated by the gravitational influence of Jupiter. Another interesting family of comets is made up of comets that come very close to the sun at perihelion. These comets may be the descendants of a single comet that broke up during a pass near the sun.

Sun-Grazing Comets

One of the orbital elements that is tabulated in Marsden's *Catalogue of Cometary Orbits* is the perihelion distance, represented

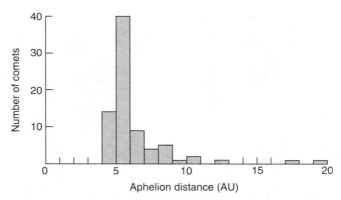

Figure 2.11 The distribution of aphelion distances for the short-period comets. Note the concentration around Jupiter's heliocentric distance of 5 astronomical units. (Data from B. G. Marsden, *Catalogue of Cometary Orbits*, 5th ed. [Cambridge, Mass.: Harvard-Smithsonian Astrophysical Observatory, 1986])

by the letter q. This element is the comet's distance from the sun when it is at its closest, and it is measured from the center of the sun. If you run down the list of values, you find comet Halley with $q = 0.587$ AU, and a great many other comets with q that ranges around 1 AU. There are some real anomalies in the table; comet 1843 I, for example, with $q = 0.0078$. Bear in mind that the smaller the value, the closer the comet passes by the sun; it may even pass below the sun's surface. A comet with $q = 0.005$ AU, for instance, passes within 750,000 kilometers of the center of the sun. That seems like a great distance until you realize that the radius of the sun is about 700,000 kilometers. The distance of comet 1882 II above the sun's surface at its closest approach was less than 20 percent of the sun's radius, a pretty close call for the comet. Comets with such small values of q are called *sun-grazing comets*.

Sun-grazing comets tend to be exceptionally bright. Comet 1843 I, seen in broad daylight in February 1843 about 4 degrees from the sun, was no exception. The comet (Figure 2.12) was then 40 times brighter than the planet Venus. The nucleus of a comet passing that close to the sun is exposed to two dangers: (1) the extreme gravitational attraction of the sun creates huge *tidal forces* (forces that are greater on one side of the nucleus than on the other) that could tear the comet apart;

Figure 2.12 Woodcut of the Great Comet of 1843 over Paris on the night of March 19. (From A. Guillemin, *The World of Comets* [London, 1877])

and (2) the intense solar energy increases the amount of radiant energy falling on the comet, which in turn increases sublimation of the nuclear ices.

When the Great Comet of 1843 was closest to the sun, it was well inside the sun's 2-million-degree *corona* (the tenuous outermost part of the sun's atmosphere, seen at total solar eclipses). It was in no danger from the high temperatures there, however. The amount of heat energy contained in a gas is determined by the total kinetic energy (that is, energy of motion) of all the ions, atoms, or molecules of the gas, whereas the temperature is a measure of the average energy of the individual particles (that is, ions, atoms, or molecules). The solar corona contains 10^8 particles per cubic centimeter. A cubic centimeter of air at room temperature contains over 10 million times more heat energy than a cubic centimeter of solar corona. The low-density corona does not contain enough heat energy to melt the comet's nucleus. The Great Comet, however, received 30,000 times more radiant energy in the form of sunlight than it would at the earth's distance from the sun.

An ideal body known as a *black body* sitting 120,000 kilometers above the sun's surface would quickly heat up to 5000°C—a temperature that could vaporize any known substance. The comet survived because it moved very rapidly, and the process of sublimation carried the heat away quickly.

We have evidence of at least one comet that did not miss the sun. The U.S. Naval Research Laboratory (NRL) in Washington, D.C., designed and built a space flight instrument to produce images of the solar corona. The objective of the instrument was to monitor long-term changes in the corona. The monitor worked successfully for a number of years—until the U.S. Air Force satellite on which it flew was used in an anti-satellite warfare test.

The instrument produces an image of the sun's outer corona from a distance of 2.5 to 10 solar radii. The light from the bright disk of the sun and the inner corona is blocked from entering the camera by an occulting disk. Figure 2.13 shows a series of images made by the camera in late August 1979. The white inner circle in each image was added artificially to represent the sun. We see a comet approaching the sun, then disappearing behind the occulting disk. Soon thereafter, the corona abruptly brightened significantly. It appears that the comet came so close to the surface of the sun (if it didn't simply plow into it) that it was unable to withstand the incredible heat and tidal forces. Apparently the entire nucleus was vaporized, and the material caused the brightening of the corona.

The individuals after whom comet Howard-Koomen-Michels was named were all NRL investigators. They measured the position of the comet on their images and calculated its orbit (Figure 2.14). The perihelion distance calculated had relatively large uncertainty, so it was unclear how close the comet actually came to the sun—but it was close enough to destroy the comet.

Comet Howard-Koomen-Michels was but one of 15 sungrazing comets detected by coronagraphs on spacecraft during the ten years from 1979 to 1989. With only one possible exception, none was detected from the ground. All met the same fate—none survived perihelion passage.

Incidentally, the elements of the orbits of all these sungrazing comets are surprisingly similar. The same similarity in a much smaller sample was noted by Heinrich Kreutz about

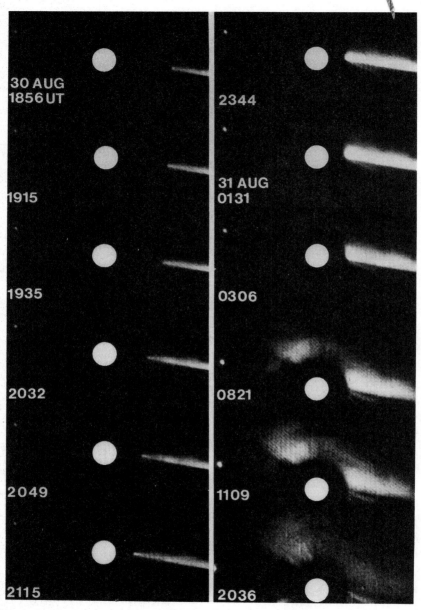

Figure 2.13 Comet Howard-Koomen-Michels hitting the sun on August 30 and 31, 1979. (Photographs courtesy of Naval Research Laboratory, Washington, D.C.)

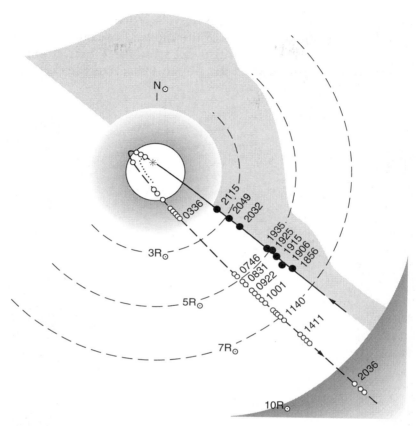

Figure 2.14 Orbit of comet Howard-Koomen-Michels near the sun. The symbol $N_\odot$ marks the north pole of the sun. The projected distances are measured in solar radii ($R_\odot$). (Courtesy of Naval Research Laboratory, Washington, D.C.)

100 years ago. Kreutz suggested that the comets form a family—now called the *Kreutz family*—that might have been formed when a much larger comet was split apart in a close encounter with the sun. Brian Marsden found that the progenitor might well have been the famous comet that Aristotle observed in 371 B.C.

The accurate prediction of the return of Halley's comet in 1759 ended once and for all any lingering doubts about the nature of comets as objects of the solar system. In the remainder of the 18th century and into the early 19th, celestial mechanics blossomed into a highly developed branch of mathematical science. As predictions of the motions of comets became more and more precise, scientists began to turn their attention toward efforts to understand the physical characteristics of these bodies.

The Heads of Comets

The Nucleus

During the late 19th century, the idea arose that the comet's center is composed of sand or gravel—that is, relatively small, solid particles ranging in size from a few millimeters to a few meters in diameter. This view was reasonable, given the discovery that the orbits of the small solid particles that produce meteors associated them with comets. This model persisted for decades and was extensively developed. In marked contrast to the prevailing view of the comet's center as a sandbank, the idea that a comet must have a more or less solid core—a *nucleus*—began to develop. Until 1986 no one had ever seen a nucleus. But in that year close-up photographs of Halley's nucleus were obtained as several spacecraft raced by the comet. So today we have a much better understanding of the nature of at least one comet's nucleus.

In the years between 1880 and 1950 researchers had to rely on indirect evidence of the nature of cometary nuclei. An obvious question was: "How big is the nucleus?" Most questions of that sort do not depend on whether the nucleus is a sandbank or one solid body (or monolithic nucleus). When we look at a comet through a telescope, we see the nucleus as a starlike point of light. Even the most powerful telescope available could not resolve the point into a measurable disk. At best, scientists could say that the nucleus was no bigger than some specific value, because if it were larger than the specific value, they could resolve it. In 1927, for example, the comet Pons-Winnecke passed within 6 million kilometers of earth, yet the nucleus remained essentially unresolved. In that case we could say that if the nucleus had been larger than 5 kilometers we could have resolved it. A scientist would express the same thing by saying that the "upper limit" to the size of the nucleus was 5 kilometers.

A few comets have orbits that extend closer to the sun than 1 astronomical unit and by chance can pass directly between the earth and the sun. Such a situation gives researchers an opportunity to look for the nucleus silhouetted against the bright disk of the sun. The Great Comet of 1882 passed between the earth and the sun, but no silhouette was observed. The observation of a dark disk against the extremely bright disk of the sun is difficult to interpret, but the absence of such an observation led the researchers to conclude that the nucleus could have been no larger than 70 kilometers.

More than a dozen known comets have grazed the surface of the sun as they passed perihelion. Comet 1843 I, for example, was only 125,000 kilometers above the sun's surface at perihelion. Yet some of these comets were not destroyed by the intense heat. Observers concluded that the nucleus must have been larger than a few meters across, because otherwise it would have been vaporized. We can tell how intense the heat became by noting that metals such as iron were vaporized by the sun's heat. The lower limit of the size of the nucleus of a sun-grazing comet is a few meters. Note that some comets, such as the Great Comet of 1882, have split into several pieces when they approached very close to the sun. This tells us something about the internal strength of the nucleus, but does not invalidate estimates of minimum size.

Another technique that has been used to estimate the sizes of nuclei is to determine the brightness of the starlike point of light. We know how far a comet is from the sun, and therefore how much sunlight falls on it. We know how far a comet is from the earth, and therefore how much of the light it reflects reaches us. The amount of light the nucleus reflects depends on its size and its reflectivity. If we make a guess about the reflectivity (say 50 percent), then we can estimate its size. Such calculations led to an estimate of 1 to 10 kilometers for the size of the cometary nuclei to which the technique was applied. As we shall see, we were probably wrong in our estimates of reflectivity. Halley's nucleus is one of the darkest bodies in the solar system, with a reflectivity of only a few percent.

Today it is generally agreed that the comet's nucleus is a single solid body. The concept of a sandbank nucleus is entirely inconsistent with current understanding, yet the most authoritative text on astronomy still presented it as the prevailing view as late as 1945.

The year 1950 marks the beginning of the modern era in cometary studies. That year Fred Whipple described the nature of the nuclear body, Ludwig Bermann described the solar-wind interaction, and Jan Oort described the reservoir of new comets.

Whipple's Dirty Snowball

A major weakness of the sandbank model of the nucleus is that it depended on the processes of adsorption and desorption to produce the comas and tails of comets. *Adsorption* is the process by which molecules of a substance are bound to the entire surface of a solid body in a layer that is roughly 1 molecule thick. The opposite process, by which adsorbed molecules are released from the surface of a solid object, is called *desorption*. Laboratory research in the 1940s established that the amount of material that can be adsorbed on a 1-gram particle is of the order of 10^{19} molecules. If the nucleus were a sandbank, all of the small particles in it would typically supply enough material to produce a coma and tail for about a week. It is simply not possible to adsorb enough material on the surfaces of the putative particles to produce the quantities of gases that have been observed. A periodic comet in particular—Halley's, say—would have desorbed (or outgassed) all its adsorbed material centuries ago and faded into obscurity. There is no reason to believe that a comet ever recovers a significant amount of the material it loses. The only effective alternative to the adsorption-desorption model is a nucleus composed entirely of the volatile material itself, which it loses slowly enough to form the coma and tail over a long period of time.

In 1950 Fred Whipple (Figure 3.1) proposed the icy-conglomerate model of the nucleus, or, in more picturesque terms, the dirty-snowball model. Whipple described the nucleus as a great mass of ice or snow with dust particles embedded in it (Figure 3.2). As a comet approaches the sun, the ices in the nucleus *sublimate*—that is, pass directly from the solid to the gaseous state—producing the gases that make up the coma and eventually the tail. As ice sublimates, the particles that are embedded in it are released into the coma as well. Only the

Figure 3.1 Fred Whipple.

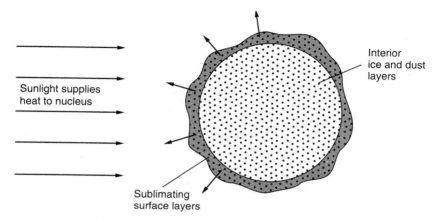

Figure 3.2 Simple schematic of a comet nucleus.

surface layers of ice sublimate; the rest of the outer layers remain to insulate the bulk of the nucleus. Large mounds of snow deposited along a roadside by a snowplow last many days for the same reason. Thus each time a comet appears, only material from its surface layers sublimates, so the comet can last for hundreds or thousands of apparitions.

Researchers were naturally curious about the composition of the ice. When a comet is roughly 3 astronomical units from the sun, bands of some molecules—cyanogen (CN), for example—appear in the spectrum. Whatever makes up the nucleus must be able to be sublimated by the relatively low solar heating (roughly 10 percent of the value at earth) at 3 astronomical units. The likely candidates were water (H_2O), methane (CH_4), ammonia (NH_3), and carbon dioxide (CO_2). The one substance that turned out to fill the bill was plain, everyday water ice (H_2O). The cyanogen molecule is just a trace cosmic pollutant in the nucleus. Why, then, do we see evidence of cyanogen in the spectrum but no evidence of water? We will have more to say about this later, but a partial answer is that evidence of the cyanogen molecule is easily accessible in the visible region of the spectrum, whereas the strong spectral bands of the water molecule occur at longer wavelengths that are not visible.

Additional support for the dirty-snowball model came from its ability to explain a vexing problem in cometary orbits, small

discrepancies thought to be caused by *nongravitational forces*—
forces resulting from nonuniform sublimation of ice on a comet's nucleus. This realization was part of Whipple's original development, but we discuss it later in this chapter in connection
with the rotation of nuclei. If we are to understand the modern model of a nucleus, we first must understand something
about the physics of water ice.

The Physics of Water Ice

Water boils at 100°C at atmospheric pressure typical of sea level.
As anyone who has tried to hard-boil an egg at a high altitude
knows, however, the lower the atmospheric pressure, the lower
the temperature at which water boils. At the low pressure of a
high altitude, the water boils and evaporates before it is hot
enough to cook the egg. On a diagram of pressure versus temperature for water, solid water (ice), liquid water, and gaseous
water (water vapor) occupy different regions. Of interest to us
here is the fact that there is a pressure (the triple-point pressure) below which liquid water cannot exist. At pressures lower
than the triple-point pressure, water goes directly from ice to
water vapor. This is the process known as sublimation. Pressures at the surface of cometary nuclei are below the triple-
point pressure, and therefore sublimation plays a major role
in cometary physics.

    Water ice has an interesting internal structure. Solid-state
physicists recognize many kinds of ice, some of which occur
only under conditions of high pressure. The common forms
and the ones of interest are amorphous, cubic, and hexagonal
ice. We can see the physics involved in the passage from one
form of ice to another by considering the way a vapor deposits
ice on a laboratory cold plate. At very low temperatures, the
water molecules stick to the cold plate in no particular order
or arrangement. The strong forces that bind individual molecules together and the low temperatures that impede their
movement prevent them from rearranging themselves into a
regular structure. This is amorphous ice. (Glass is a common
amorphous solid.)

    If the plate is heated, the molecules can partially over-
come the strong forces (bonds) and rearrange themselves into

a lattice with molecules on the vertices of cubes; hence "cubic ice." Because the configuration of cubic ice has a lower energy than that of amorphous ice, the transformation occurs with a release of energy; this transition is said to be exothermic, and it occurs at approximately 136 degrees Kelvin (K). (The Kelvin temperature scale has the same size degree as the Celsius scale, but uses absolute zero as its zero point. On the Kelvin scale, 0 K = −273°C.)

If the plate continues to be heated, the cubic ice changes into a form in which the basic element of the crystal is a hexagon. The hexagonal configuration is at a lower energy than the cubic configuration, and this transition is also exothermic. It occurs at approximately 173 K. Ordinary refrigerator ice is hexagonal ice.

Linus Pauling won the Nobel Prize for his discussion of the hydrogen bond, a strong force that works at the molecular level and attracts water molecules to one another. When water freezes, it forms into a crystal structure, the molecules held together by that hydrogen bond. The structure resembles a cage, with large cavities within it (Figure 3.3). This loose struc-

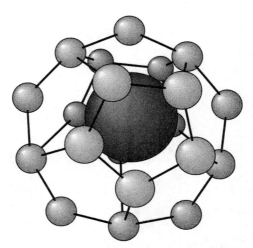

Figure 3.3 A clathrate hydrate. The outer spheres, which form a cage, represent one unit of a water ice crystal. When molecules are trapped in the cage, the clathrate hydrate is formed. Here the trapped molecule (darker shading, center) might be a molecule of carbon dioxide.

ture, incidentally, explains why ice is less dense than water and therefore floats. The bonds that hold the ice together can produce an interesting substance called *clathrate hydrate*. If ice condenses in the presence of other gases, some "impurities" can be trapped in the cages in the crystal lattice. Up to 1 part impurities to $5\frac{3}{4}$ parts water can be locked up.

The Nucleus Model Refined

The picture of the nature of ices in the nucleus has been refined by several researchers, chief among them Armand Delsemme (Figure 3.4) and Pol Swings. They hypothesized that

Figure 3.4 Armand Delsemme.

such substances as methane and ammonia, as well as other atoms and molecules, are trapped in the cavities of a clathrate hydrate. When the water ice sublimates, the imprisoned substances are released. These impurities, such as molecules of carbon (C_2), carbon monoxide (CO), and cyanogen (CN) in comets, are released when the lattice disintegrates, an event that is determined by the thermodynamic properties of water. This property accounts for certain observations of cometary spectra.

Later, when we talk about the origin of comets, we will present evidence suggesting that other organic molecules, such as formaldehyde (H_2CO), isocyanic acid (HNCO), and formic acid (HCOOH), are imprisoned in nuclear ice. These molecules are difficult to observe, however. As soon as they are released into space, ultraviolet photons from the sun will dissociate them—that is, tear them apart—into simple elements or structures, such as cyanogen and hydroxyl (OH). This process is known as photodissociation, and the pieces resulting from it are called daughter molecules. The task facing students of comets is to identify the parent molecules from which these daughters arise. We will have more to say about this problem later, when we talk about past and future space missions to study comets.

The picture of the nucleus that has developed since 1950 is one of a large mass of water snow with small particles and trace substances embedded in it. This picture explains a great deal of the observed cometary phenomena that we will talk about in the following pages. Later we will describe how the intense research activity associated with the return of Halley's comet in 1986 has refined, extended, and verified our picture of comets. This model has stood the test of time. For this and other contributions to cometary science, Fred Whipple was awarded the American Astronomical Society's prestigious Henry Norris Russell Prize in 1987.

Modern Models of the Nucleus

Fred Whipple's icy-conglomerate model provides the foundation of modern comet theory. In essence, the nucleus is envisioned as a dirty snowball, made of water ice and dirt provided by dust particles believed to be distributed throughout the body. When the comet's orbit takes it far from the sun, the nucleus is very cold and essentially inert. As the comet approaches the sun, the sun's radiant energy (sunlight) heats the surface. As it recedes from the sun, the surface of the nucleus has gained energy from sunlight but now loses energy by black-body radiation to strike a balance, if we overlook the small amount of energy conducted into the interior of the nucleus. As the comet nears the sun again, the solar radiant input exceeds the nucleus's radiant output, and the temperature of the surface increases. When the temperature rises to values determined by the thermodynamic properties of the ice, sublimation begins. Over a range of temperatures or a range of heliocentric distances—it amounts to the same thing—the surface continues to gain energy from sunlight but loses energy by two processes: black-body radiation and sublimation of ices. Finally, when the comet is sufficiently close to the sun, essentially all the input energy from sunlight goes into sublimation, and the temperature no longer increases. The various energy-balance regimes are seen in Figure 3.5.

The sublimation of the ices produces a porous dust crust, which tends to insulate the ices beneath and regulates the rate of sublimation now taking place a few centimeters below the surface. Energy is conducted (or otherwise transported) down to the ices, which sublimate, and the gases percolate through the porous dust layer to escape. The rate of sublimation and crust formation need not be uniform, either over time or over the surface of the nucleus; thus enhanced sublimation (jets) and, over time, surface features and an irregular nuclear shape are produced. We will see later that the jets appear to be active only on the sunward side, a fact that implies a thin crust in these areas. The fraction of the surface involved in sublimation is believed to vary from 100 percent for a young comet to an intermediate value of about 10 percent for an older comet such as Halley to values approaching zero for extinct comets.

A cometary coma generally appears at a heliocentric distance of about 3 astronomical units, which is consistent with the thermodynamic properties of water in combination with the sublimation process. Water ice is likely to be the principal constituent at least of the outer layers, particularly in light of the fact that water-group molecules such as H_2O, hydroxyl

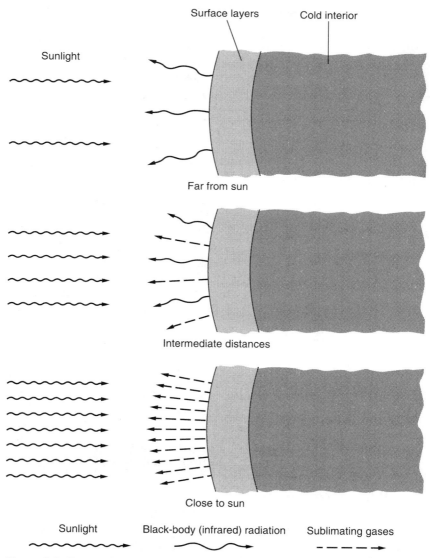

Sunlight

Surface layers   Cold interior

Far from sun

Intermediate distances

Close to sun

Sunlight   Black-body (infrared) radiation   Sublimating gases

Figure 3.5 Energy-balance regimes for cometary surface layers at three distances from the sun.

(OH), the hydroxyl molecular ion (OH^+), the water ion (H_2O^+), and the hydronium ion (H_3O^+) predominate in the coma.

The dominance of water ice in the nucleus, however, may not apply to the composition of the total bulk of the nucleus or to a comet approaching the sun for the first time. Before we explain this idea, we need to deal with another observation—that bands of cyanogen (CN) and molecular carbon (C_2) appear in cometary spectra when the coma first forms near 3 astronomical units. Almost all substances other than water should begin to sublimate much farther from the sun than 3 astronomical units. The solution to this problem was to postulate that the water ice occurs as a clathrate hydrate. The thermodynamic properties of the clathrate hydrate are normally close to the thermodynamic properties of pure ice. Thus the clathrate hydrate sublimates at the expected distance from the sun and releases the trapped minor constituents. The cyanogen and molecular carbon molecules are detected because they are spectroscopically favorable; that is, their bands are in a spectral region accessible from the ground, and the physical processes that produce the bands occur at the conditions encountered in the coma at 3 astronomical units.

Now let us return to the dominance of water ice. In terms of total composition, consider the ratio of water to carbon dioxide, which has the value 80:3.5 in Halley's comet. When observations of comets close to the sun are made, this ratio is much lower. The discrepancy is easily explained if the composition of the surface layers is not representative of that of the entire comet. The heating of the comet by sunlight can be considered as a thermal wave traveling inward. When a comet passes near the sun, the wave travels deep and causes sublimation of the pristine ices. The composition of the ices deep in a comet may be different from the composition of the surface layers of a comet that has made many passes through the inner solar system.

When a comet approaches the sun for the first time, an interesting situation can occur. The capacity of the clathrate hydrate to store minor constituents is approximately 17 percent of the number of water molecules. Thus a minor constituent with an abundance of more than 17 percent of the water ice abundance cannot be completely stored in the lattice. The sublimation of these excess minor constituents is controlled by their own thermodynamic properties; it generally occurs at distances greater than 3 astronomical units. Thus fresh comets may be active or brighter than normal at great distances.

The specific abundance of water in cometary nuclei is uncertain, and in some circumstances other substances can be important in the sublimation process. Nevertheless, water ice is still the substance that has the greatest influence on the behavior of most comets.

Substances other than water and carbon dioxide are also significant in comets. We have already mentioned cyanogen and molecular carbon; the molecules carbon monoxide (CO) and polymerized formaldehyde $(H_2CO)_n$ have been observed, and methane (CH_4) and ammonia (NH_3) are expected. The carbon monoxide molecule could be the source of the ionized carbon monoxide (CO^+) that dominates photographs of plasma tails in color or blue-sensitive emulsions. The presence of polymerized formaldehyde (either alone or as an indicator of other complex organic compounds that cannot be observed directly) may explain the exceptional darkness of a comet's surface. The surface material may be intrinsically dark, and multiple reflection by the rough surface of the nucleus will certainly decrease the reflected light. A layer of polymerized formaldehyde on the crust material is very dark and would contribute to the remarkably low *albedo*—the fraction of light that is reflected by the surface. The polymerized formaldehyde could have been produced either long ago by the same processes that formed the dust grains or now by exposure to ultraviolet light.

A dark surface should be close to the temperature of a black body at the same heliocentric distance. The temperature of the outer dust crust of a comet near the orbit of earth should be approximately 300 K. Energy is transferred from the outer crust down to the layer containing the sublimating ice, which has a temperature of approximately 215 K. The temperature of the deep interior could be 50 K or lower.

Figure 3.6 shows the nucleus of a comet. Details of the comet's interior structure are suggested in Figure 3.7. Our knowledge of cometary interiors is meager. We do not know, for example, if a large nucleus such as Halley's is truly monolithic or is composed of *cometesimals*—smaller pieces that are thought to make up the nucleus—say, 0.5 kilometer in diameter. Because rotation is certainly a universal property of cometary nuclei, the nongravitational forces can be generated by a mechanism we shall describe shortly, and which requires rotation. But, as we shall see, this explanation has been challenged recently.

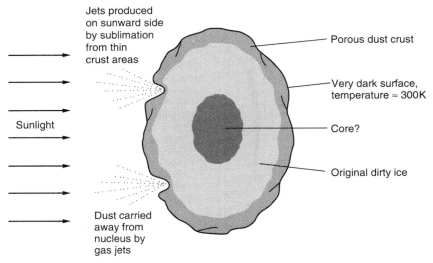

Figure 3.6 Schematic of a comet nucleus showing some complexity and the possibility of a central core.

The masses and densities of nuclei are still somewhat uncertain. The traditional method of estimating mass is simply to assume a density appropriate to a large chunk of ice (usually 1 gram per cubic centimeter) and to use dimensions estimated by one of several methods. For a sphere of radius r, the mass is approximately $4r^3$. Because the masses of some comets (such as Halley) are believed to be known from studies of the non-gravitational forces, and because the linear dimensions of Halley's nucleus were larger than anticipated, many comet scientists believe that the density is low, perhaps as low as 0.2 gram per cubic centimeter. This value may be correct, but it is quite uncertain. The density of the overall nucleus should be in the range of 0.2 to 2.0 grams per cubic centimeter, with a slightly preferred value near 0.5. With this assumption, Halley's comet currently has a mass of about 2×10^{18} grams.

The Coma

The coma and the nucleus together make up the comet's head. The *coma* is a more or less spherical cloud of dust and neutral molecules surrounding the nucleus. As Figure 3.8 shows, the

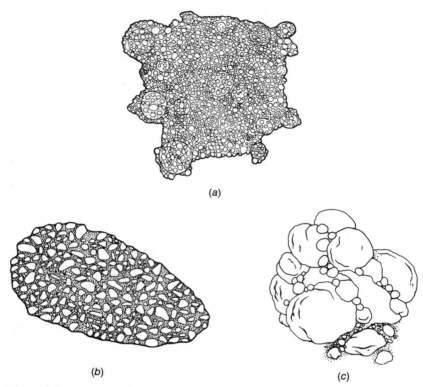

(a)

(b)   (c)

Figure 3.7 Possible models of the interior of a comet's nucleus. *(a)* Cross-sectional view of an aggregate model. Cometesimals (indicated by circles) are aggregates of micron-sized ice-dust particles. Some cometesimal identity is retained in the nucleus. (After B. Donn and J. Rahe) *(b)* Cross-sectional view of an icy-glue model. Boulders are held together with a glue of a Whipple-type dust-and-ice mixture. (After T. I. Gombosi and H. L. F. Houpis) *(c)* Perspective view of a rubble-pile model. The nucleus consists of smaller fragments bonded together by local melting at contact surfaces. The interior could resemble *a* or *b*, or it could have a solid core (as in Figure 3.6). (After P. R. Weissman)

size of the coma appears in photographs to increase as the exposure time increases. This tells us that the brightness of the coma decreases as distance from the nucleus increases. Longer exposures bring out the increasingly fainter outer layers of the coma. On a planet the coma would be considered simply a dusty atmosphere.

Since a typical coma has no obvious outer edge, it is difficult to say how big it is. We might define the size as the dis-

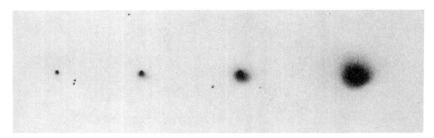

Figure 3.8 The coma of Halley's comet on May 25, 1910. The exposure times are (left to right) 15 seconds, 30 seconds, 1 minute, and 3 minutes. (From J. Rahe, B. Donn, and K. Wurm, *Atlas of Cometary Forms* [NASA SP-198, 1969])

tance from the nucleus to the point where the brightness is 50 percent, 10 percent, or some other percentage of its value at the nucleus. The radius of a coma tends to be about 10^5 kilometers, but varies between 10^5 and 10^6 kilometers. We cannot simply ascribe a certain size to the coma of a given comet, however, because the size has been found to depend on the comet's distance from the sun. When comet Halley was first seen in October 1982 on its way to its most recent pass by the sun, it was farther away than the planet Saturn. All that was available to reflect light was the comet's solid nucleus. Typically a coma begins to become noticeable when a comet is 3 or 4 astronomical units from the sun. As the comet approaches closer to the sun, its coma continues to grow. It reaches its maximum size when the comet is between 1.5 and 2.0 astronomical units away. As the comet approaches yet closer to the sun, the coma usually shrinks. The material in the coma comes from the nucleus by the sublimation process, so one might expect that process to be stimulated by the heat of the sun, and therefore the coma should continue to grow. But other processes at work ultimately remove material from the coma, sending some of it into the tail. At some point the removal process overwhelms the creation process as a comet approaches the sun.

The Hydrogen-Hydroxyl Cloud

In 1970 technicians at NASA directed the ultraviolet detectors of the second *Orbiting Astronomical Observatory (OAO-2)* toward comet Tago-Sato-Kosaka. The resulting observation revealed a cloud of hydrogen much larger than the comet's coma (actually about 0.1 AU in diameter) surrounding the comet. Later similar clouds were discovered around other comets as well. Further studies of the ultraviolet spectra of these comets from space also revealed the presence of a somewhat smaller hydroxyl (OH) cloud. These two clouds are now known as the *hydrogen-hydroxyl cloud* (see Figure 3.9 and Plate 5).

It is interesting to note that this cloud complex is made up of the dissociation products of water. That is, if ultraviolet photons in sunlight were to break up a water molecule (HOH), the products would be a hydrogen atom (monatomic hydrogen, H) and a hydroxyl *radical* (OH). The breakup of hydroxyl also produces monatomic hydrogen. Thus a sublimating body

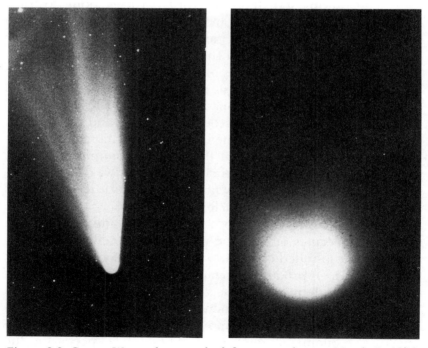

Figure 3.9 Comet West, photographed from a rocket on March 5, 1976. Left: Visual light image. (Courtesy of P. D. Feldman, Johns Hopkins University) Right: Ultraviolet image in Lyman-alpha showing the hydrogen cloud. (Courtesy of C. B. Opal, University of Texas, Austin, and G. B. Carruthers, Naval Research Laboratory, Washington, D.C.) The two images are at the same scale.

composed largely of water ice will produce water vapor, which is dissociated by sunlight to yield the huge cloud of hydrogen. The idea of a hydrogen-hydroxyl cloud around comets was developed by Ludwig Biermann in 1968 from a suggestion that he and Eleonore Trefftz had advanced in 1964. In retrospect, the idea was very straightforward.

Models of the Coma and the Hydrogen Cloud

The neutral molecules produced by sublimation constitute the gaseous part of the coma. The gravitational attraction of the nucleus is very small and cannot hold the gas. Therefore, the gas streams off in a flow, physically analogous to the *solar wind,* the plasma flowing off the sun. The flow reaches a final speed or terminal bulk speed of about 0.5 kilometer per second. The flowing gas drags the dust particles along to form the dust component of the coma. When the ices sublimate, the freshly liberated porous dust particles can be left behind either as loose particles or as part of the dust crust. The gas flow then carries away the loose particles or strips additional particles from the edges of the crust. The gas near the nuclei of most comets is dense enough that collisions between molecules are important. Chemical (gas-phase) reactions can occur, and the breakup of molecules by sunlight (called photodissociation) can be important, too. This means that the relatively simple molecules, such as carbon monoxide and cyanogen, that we observe spectroscopically in the coma may not be the same as the molecules initially released from the nucleus.

The hydrogen cloud is produced by the photodissociation of water molecules outside the region where collisions are important, the so-called collision zone. Photodissociation of water molecules (H_2O) produces monatomic hydrogen atoms (H) with speeds of 20 kilometers per second and slow-moving hydroxyl molecules (OH); the subsequent photodissociation of the hydroxyl molecules produces monatomic hydrogen atoms with speeds of 8 kilometers per second. These great speeds and the absence of collisions enable the hydrogen atoms to travel long distances before they are ionized by solar radiation or the solar wind. The result is the large hydrogen cloud surrounding the comet.

The molecules and dust in the coma and the atoms in the hydrogen cloud are subject to the sun's radiation pressure. The effects are the most pronounced on the hydrogen cloud, which is measurably distorted.

Do Cometary Nuclei Rotate?

If cometary nuclei are multikilometer-size solid objects, do they show any evidence of rotation? This question arises because rotation is a common property of celestial bodies—asteroids, satellites, planets, stars, and galaxies—and the presence of rotation in cometary nuclei would have important consequences. The answer is an unequivocal yes. One of the most direct bits of evidence for rotation is the spiral structures in the coma. The model assumes that jets of dusty material are released from the nucleus, then drawn into spirals by its rotation. Comet Bennett (1970 II) (Figure 3.10) showed a well-developed spiral structure in its coma, which suggested a nuclear rotation period of 34 hours. Less direct evidence for nuclear rotation comes from so-called nongravitational forces.

A number of comets—among them comets Halley, Encke, and Pons-Winnecke—show what are called *secular accelerations*. The mathematician Johann Franz Encke corrected the motion of Encke's comet for the effects of every known planetary perturbation and found that the orbital period was still shortening by roughly 0.1 day each revolution. One possible explanation was that material between the planets was resisting the comet's motion, slowing it down. The forces at work came to be called *nongravitational forces* because they cannot be explained by any known gravitational influence in the solar system.

In 1836 Friedrich Wilhelm Bessel suggested that the jets of material emitted by nuclei could act as rockets to either accelerate or decelerate the nucleus. But there is a problem with this idea. Since sunlight falls on the side of the nucleus toward the sun, the rocket effect would be in the radial direction and thus perpendicular to the direction of the comet's motion. If a nucleus does not rotate, the rocket effect essentially changes the sun's gravitational attraction, and consequently the comet's orbit shrinks or expands. The traditional explanation, how-

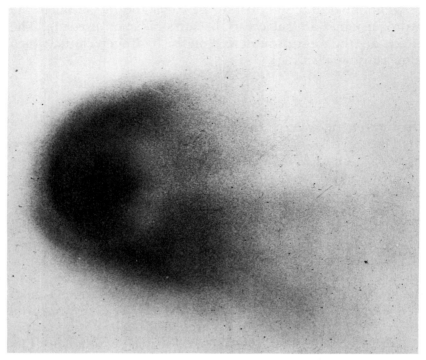

Figure 3.10 The spiral structure in the head region of comet Bennett is visible in this photo taken March 20, 1970. (Courtesy of S. M. Larson, Lunar and Planetary Laboratory, University of Arizona)

ever, is that only forces with a component in the direction of the comet's motion can accelerate or decelerate it.

The answer was provided by the rotation of the nucleus. The point that is heated by sunlight will always move a short distance because of rotation during the time when the sublimation takes place (Figure 3.11), and the rocket effect will not necessarily be radial. Depending on the direction of rotation, the rocket effect can accelerate or decelerate the nucleus. As we mentioned earlier, this explanation of the nongravitational force was an integral part of Whipple's development of the dirty-snowball model. The process has been studied in detail by Zdenek Sekanina, and it is now generally accepted. One final note: if the rotation rate were very rapid, it could smear out the rocket effect. Thus it is clear that some cometary nu-

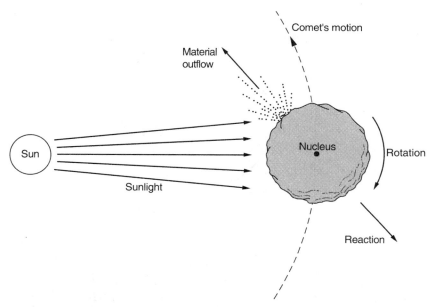

Figure 3.11 The origin of nongravitational forces on the nucleus of a comet. Reaction to material outflow is a force with a component opposite to the comet's motion.

clei must rotate relatively slowly. The average period of all rotations calculated up to about 1980 was 23 hours.

The traditional explanation has been challenged by Donald K. Yeomans and Paul W. Chodas. If the rate at which gas is produced is not the same before and after perihelion (usually it is greater after perihelion), the change alone (that is, without rotation) can cause the orbit to shrink or expand and can produce the effects usually ascribed to the mechanism involving rotation.

Split Comets

The Austrian astronomer Wilhelm von Biela discovered comet 1826 I in February 1826. The comet remained visible for several months, and Biela collected enough positional observations to permit accurate calculations of the elements of its or-

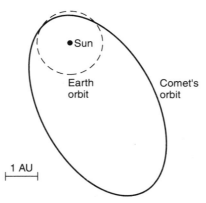

Figure 3.12 The orbit of comet Biela (1832 III) projected onto the plane of the earth's orbit.

bit. A comparison of these orbital elements with those of earlier comets clearly showed that the comet had been seen before. It had been observed in 1772 by Montaigne and in 1805 by Jean-Louis Pons, but the earlier observations were insufficient to allow the calculation of accurate orbital elements. The comet was named Biela's comet.

Biela's comet had a period of only 6.6 years; the orbit is shown in Figure 3.12. The comet was recovered (that is, seen again on this pass through the inner solar system) in 1832, approximately on schedule. The comet was expected to reach perihelion in the summer of 1839, when its perihelion point and the earth were on opposite sides of the sun. The comet remained lost in the sun's glare and was not seen on that pass. It was seen in 1846, not as the expected single comet but surprisingly as two comets, close together in the sky. Biela's comet had become a *split comet*. Both comets had tails (Figure 3.13). During the time the comets were visible they gave an extraordinary show; first one, then the other grew brighter. In addition, the comets showed unusual activity. One grew a second tail, and at times a luminous bridge of material joined the two comets, which remained about 250,000 kilometers apart.

In 1852 both comets returned on schedule, about 2 million kilometers apart. The same sort of behavior was observed on this pass. The two comets took turns being the brighter of the pair, and a luminous bridge was seen between them. On the next pass, Biela was once again lost in the glare of the sun

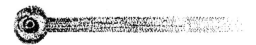

Figure 3.13 The components of comet Biela on February 19, 1846. (Drawing by O. Struve. Reprinted by permission of the Smithsonian Institution Press from Fred L. Whipple, *The Mystery of Comets,* p. 82. Copyright © Smithsonian Institution, Washington, D.C., 1985.)

and was not seen. The 1866 appearance was expected to be very favorable for observation, but the comet did not return.

There is some controversy as to whether Biela's comet has ever been seen again. In 1872 the earth crossed the orbital path of the comet and encountered an unusually strong meteor shower. The German astronomer Ernst Klinkerfues concluded that the meteor shower was caused by the earth's passage close to the comet. Klinkerfues sent a telegram to a colleague at the Madras Observatory in India, urging him to search for the comet in an area of the sky where it should have been as it receded. The colleague, N. R. Pogson, found a comet near the expected point and observed it for several days, but it was probably not Biela's comet. We are reasonably sure that Biela's comet was never seen again.

Other comets besides Biela have been seen to split; one did so in 1976. Observations show that the nucleus of comet West (1976 VI) split into four components (Figure 3.14). They remained so close together that rather than four distinct comets, they formed a single comet with four nuclei. Such splitting is not at all unusual. One aspect of the buildup of bodies in the solar system—whether planets or cometary nuclei—may be directly relevant to comet splitting: the collision and coalescing of smaller bodies to make larger ones. The cratered surfaces of the moon, Mercury, and Mars offer ample evidence of many

Figure 3.14 Nuclear splitting in comet West (1976 VI). The dates shown (left to right) are March 8, 12, 14, 18, and 24, 1976. (New Mexico State University Observatory)

enormous collisions early in the history of the solar system. Gravitation and internal heat homogenized the structures of the colliding bodies to form the planets, but cometary nuclei are too small to be affected by those mechanisms.

One suggestion is that the larger cometesimals may not be tightly bound together, and under some circumstances, not yet well understood, could come apart. We know that nuclei rotate, and the rotation generates inertial forces that attempt to tear the nucleus apart. As the nuclear ices sublimate, the connections between the cometesimals weaken (perhaps thermal stresses are important) and finally break. If this picture is correct, then breakup could be a common event in the later stages of the life of a cometary nucleus.

Outbursts

Outbursts have been observed in many, many comet. An *outburst* is a rapid increase in brightness by as much as a factor of 100 that may last as long as several weeks, then gradually subside. Comet Schwassmann-Wachmann I—a periodic comet with a period of 15 years—has had nearly 100 outbursts (Figure 3.15) since it was discovered in the 1920s. At its most intense outburst its brightness increased by a factor of more than 1000.

One might expect that outbursts would be most common when comets are near the sun and exposed to intense heat, but that expectation is not borne out by the facts. Comet Schwassmann-Wachmann I, for example, never gets closer to the sun than 5 astronomical units. Other comets, too, have been observed to be active when they are relatively far from the sun.

Oct. 12   Oct. 18   Nov. 3

Figure 3.15 Periodic comet Schwassmann-Wachmann 1 in 1961. An outburst is clearly visible. (Courtesy of E. Roemer, University of Arizona; official U.S. Navy photograph)

A shell of material has been observed expanding away from the nucleus at speeds of several hundred meters per second in comets that have just experienced an outburst. It appears that the process leading to the outburst explosively increases the amount of dust released from the nucleus. We have no idea what that process is. One of the many explanations suggested is that there are large cavities within the nucleus where pockets of gas build up over time. Eventually the gas pressure could become great enough to rupture the pocket, spewing out dust along with the released gas.

Comet Halley has set something of a record for observed outbursts. In February 1991, when the comet was between the orbits of Saturn and Uranus, astronomers at the European Southern Observatory were making routine monitoring observations of it when it suddenly increased in brightness by a factor of more than 300 and expelled a dust cloud that reached about 300,000 kilometers in diameter. The startling observation was verified by astronomers at the University of Hawaii's Institute of Astronomy.

Regular Changes with Varying Distance from the Sun

Even a normal comet with no spectacular behavior such as outbursts or splitting displays changes as it passes through the inner solar system. As the cold and inert nucleus approaches the sun, solar radiation heats the surface layers and activity begins. The activity can be observed at approximately 3 astronomical units from the sun, but some activity can occur at greater distances. At such distances the coma forms, and cyanogen and molecular carbon bands appear in the spectrum. By a heliocentric distance of 1.5 astronomical units a plasma tail is usually present.

The brightness of the comet shows large short-term variations. They are observed by photometry, and we know from comet Halley that the gas and dust jets are produced by sunlight; hence, as the nucleus rotates about its axis, the jets rapidly "turn on" at sunrise and "turn off" at sunset. If these short-term fluctuations are ignored, comets approaching the sun *on average* brighten as the inverse fourth power (r^{-4}) of the heliocentric distance. A comet following this brightness law should be $(1.5)^4 = 5$ times brighter at 1 astronomical unit than at 1.5.

To complicate the issue, comets are often much brighter postperihelion (after their closest approach to the sun) than preperihelion (before their closest approach to the sun). Apparently there is some mechanism that stores energy and later releases it to produce this phenomenon in some comets. Comet Halley displayed this phenomenon strongly (as we shall see in Chapter 6). Yet some comets, such as Encke, are brighter preperihelion. We do not know why.

The Oort Cloud

The dirty-snowball model of the nucleus presented one thorny problem in light of all the phenomena that deplete the material in a nucleus. If you add up all the material that is observed in the coma and tail of a comet such as Halley in one complete pass by the sun, you find that the nucleus must have lost several meters of its diameter at each pass to produce the observed coma and tail. At that rate, comet Halley would dissipate a 10-kilometer nucleus in much less than a million years. If, as is currently believed, comets were formed over 4 billion years ago, when the solar system was formed, why do we still see Halley's comet? The Dutch astronomer Jan Oort (Figure 3.16) of the Leiden Observatory came to the rescue.

Figure 3.16 Jan Oort.

If we work backward and correct the observed motions of long-period comets for the perturbations of the planets, we find that their aphelion distance—the distance at which a comet is farthest from the sun—is somewhere between 50,000 and 150,000 astronomical units. Oort suggested that these long-period comets dropped out of a great cloud of comets that exists between 50,000 and 150,000 astronomical units from the sun. The outer edge of the cloud is halfway to the nearest star. Oort showed that every 10 million years or so a star with up to 1.4 times the sun's mass will wander into this cloud and come within 50,000 astronomical units of the sun. When this happens, the intense gravity of the star will affect a large number of the fledgling comets in the cloud. Some will be sent out into interstellar space, never to be seen again. But a few will be sent in toward an encounter with the sun. Some of these comets will be perturbed slightly by the massive planets when they are near the sun, and after a number of passes may become short-period comets. If there is a total of between 100 billion (10^{12}) and a trillion (10^{13}) comets in the cloud, the Oort hypothesis accounts for the observed number of long-period comets.

Oort's theory has been refined since it was originally proposed, and it continues to be a topic of discussion and research. We will have much more to say about the Oort cloud in Chapter 7.

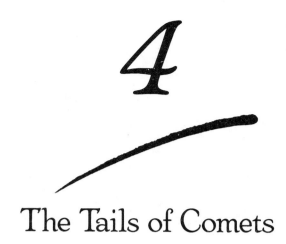

The Tails of Comets

Early Ideas about Comets' Tails

A clue to the nature of the physics of comets is something that has been known since ancient times: their tails always point away from the sun (Figure 4.1). When a comet passes perihelion and then begins its travels out into the cold nether regions of space, it actually travels tail first. In the 19th century, observers studying comets noted active structures near their nuclei (Figure 1.4). These active structures were in a constant, slow state of change and frequently seemed to curve toward the tail. It appeared as if fountains of material were being released from the sunward side of the nucleus, then were being propelled away from the sun by an as yet unknown force, which came to be called the *repulsive force*. Two 18th-century researchers, Friedrich Bessel (1784–1846) and Fedor Bredichin (1831–1904), devised a theory to explain the shapes of comet tails under the mutual influence of solar gravitation and a variety of repulsive forces.

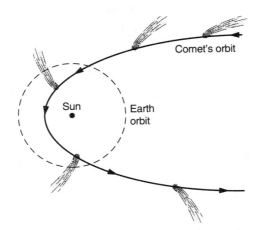

Figure 4.1 Comet tails always point away from the sun.

Careful observations of comet tails reveal that they are far from uniform. In the 18th century three types were described, on the basis of their curvature. Today we recognize only two types, which for the moment we will call Type I and Type II. *Type I tails* are straight and bluish and have a turbulent appearance like swirling cigarette smoke; *Type II tails* are curved, whitish, and relatively smooth, but occasionally have linear structures (Figure 4.2; also see Plate 6).

Two candidates were suggested for the repulsive force that pushes the material in comet tails away from the sun. A logical choice is certainly the electrostatic force, since we know that like charges repel each other. If this were the force at work, however, the sun would have to have a net electric charge that we could detect by other means. All the scientific evidence available to us today says that the sun is electrically neutral. In 1900 the Swedish chemist Svante Arrhenius (1859–1927) suggested radiation pressure as the repulsive force, and this suggestion has been verified.

To understand radiation pressure, we have to know something about the concept of particles of light, or *photons*. A photon has a tiny amount of momentum, and when it reflects off a material object it can exert a microscopic force on the object. In everyday situations we never notice the effect of that force because it is so incredibly tiny. But a particle of matter a micron (millionth of a meter) or so in diameter will be given a

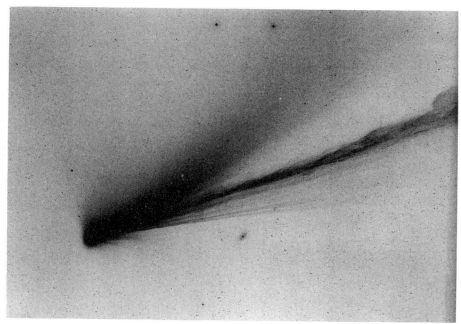

Figure 4.2 Comet Mrkos on August 24, 1957, has both Type I and Type II tails. The upper, diffuse tail is the Type II or dust tail. The lower, straight, structured tail is the Type I or plasma tail. (Palomar Observatory photograph)

noticeable push if it is hit by a photon. Could it be that comets' tails are composed of tiny dust particles that are shoved away from the sun by the impact of photons in sunlight? The answer is yes. Type II tails are composed of dust and are made to point away from the sun by radiation pressure.

Models of the Dust Tail

Dust particles roughly a micron in diameter are subjected to a strong force from solar radiation pressure and are blown in the direction away from the sun to form the dust tail. The modern theory of the process has been developed by Michael Finson and Ronald Probstein, and we find that it agrees with the shapes and sizes of dust tails when we vary the size of the particles and their rate of emission. Often a peak in dust emission before perihelion is required to make the theory and observations agree. The smaller particles form the main dust tail, and the larger particles (subject to less acceleration by radiation pressure) stay closer to the comet to form features such as the *antitail.* Under certain circumstances, dust particles can concentrate in the central area of the dust tail to form a narrow, stable feature called a *neckline* structure.

Note that radiation pressure does not explain why Type I tails point away from the sun. Understanding of this phenomenon would have to await the discovery of the solar wind that causes it and the development of plasma physics.

The Physics of Plasmas

A *plasma* is produced when the atoms or molecules in a gas have one or more electrons removed, producing a mixture of positively charged ions and negatively charged electrons. The charged particles in a plasma interact with one another and with any magnetic or electric field that may be present. Often plasmas are considered a "fourth state of matter" (the other three being solid, liquid, and gas), because plasma behavior is unique. Plasmas exhibit collective properties; that is, we often have to take a global view of a plasma if we are to understand local phenomena. Plasma behavior is strongly influenced by electric and magnetic fields. Plasma phenomena are very important in the study of the cosmos. Estimates vary somewhat, but the proportion of matter in the universe that is in the plasma state is well over 90 percent.

Physicists recognize that plasma behavior occurs when the following conditions are met: (1) The *Debye length* is small in comparison with the size of the volume occupied by the plasma. A macroscopic volume of plasma is electrically neutral—that is, there are as many positive charges as there are negative charges in the volume. If you could place a tiny probe into a plasma, you would find that at some distances from one of the many charged particles in the plasma you could not detect the charge on the particle because it is shielded by all the other charged particles. There is a distance at which the shielding is not effective, however, and you can detect the charge on the particle. This distance is called the Debye length. (2) A sphere with a radius of one Debye length contains many electrons. (3) The plasma is neutral; that is, the numbers of positively and negatively charged particles are equal, and thus there is no net charge per unit volume. (4) Plasma oscillations are not strongly damped (that is, decreased in amplitude) by collisions with ions, atoms, or molecules. These conditions are easily met in the plasma tails of comets (see the quantitative discussion in Appendix A).

No   A comet's charge is surely neutral because if there were volumes containing excess charge, large forces would be generated by the resulting electric fields, and there is little (except the magnetic field) to prevent charged particles from moving to neutralize any imbalance of charge. Plasma oscillations are not damped because the frequency of collisions is much smaller than the plasma frequency, the natural frequency with which electrons oscillate collectively around the much more massive ions.

The magnetic field in comets can restrict the movement of particles perpendicular to it (they can still move freely along the magnetic field). A charged particle moves around the magnetic field in a spiral with a radius that depends on the speed and mass of the particle and the strength of the magnetic field. The radius of an electron's spiral is typically 0.1 kilometer and that of an ion's spiral is 25 kilometers. The values are much smaller than those characteristic of dimensions in comets. Thus the plasma in comets can be regarded as highly magnetized; in other words, the magnetic field is said to be "frozen into" the plasma and they move together.

Disturbances in magnetized plasmas travel at a speed called the *Alfvén speed,* which is determined by the strength of the magnetic field and the density of the plasma. It is analogous to the speed of sound in a nonmagnetized gas. The Alfvén speed, typically 10 kilometers per second in comets, is the speed at which disturbances carry energy or information in the plasma.

The subject of plasma physics crops up repeatedly in any discussion of the physics of comets. The cometary environment provides a low-density magnetized plasma. Subjects for study include ion pickup, plasma instabilities, plasma waves, formation of current sheets, generation of high-energy particles, and collisionless shock formation. In fact, comets provide a superb opportunity to study plasma phenomena on a scale not possible in the laboratory. In comets we can probe the microscopic properties of the plasma with instruments on spacecraft and record the macroscopic properties through imaging. The combination provides a powerful basis for theoretical studies.[1]

The Solar Wind and Comets

The idea that the sun emits particles has been around at least since the turn of the century. The particles were postulated to explain *auroras* (the northern lights) and magnetic disturbances at the earth, and were thought to be concentrated largely in beams. In 1951 the German astronomer Ludwig Biermann (Figure 4.3) announced his hypothesis that there is a *solar wind*, a flow of ionized gas outward from the sun through the solar system, not in beams but in all directions. Biermann's hypothesis was based on (1) analyses of the accelerations of observed knots in plasma tails and (2) the orientations of plasma tails. If the changing positions of knots in comet tails are interpreted as motions of materials rather than as a moving wave, there has to be an acceleration mechanism. The hypothesized solar wind was the candidate. Type I tails are quite straight and point away from the sun, and spectroscopy shows that their emission comes from molecular ions, an indication that the tails are composed of plasma. A color picture of a comet shows its Type I tail to be bluish (see Plates 6, 7, and 8). The blue color arises from emission of ionized carbon monoxide (CO^+).

Detailed calculations of the interaction of the solar wind with the cometary plasma could not reproduce the measured accelerations of plasma features (such as knots and kinks) in the tail. It quickly became evident that the problem was more complex than anyone had realized.

Soon after Biermann hypothesized the solar wind, Eugene Parker of the University of Chicago showed theoretically that the ultrahot solar corona could not be a static structure but

---

1. In these brief paragraphs we have scanned a very complex field of physics. For more information see some of the references included in the Suggested Readings.

Figure 4.3 Ludwig Biermann (1907–1986).

must be continuously expanding away from the sun. His mathematical theory also showed that a high-speed solar wind should exist. In the late 1950s and early 1960s the United States and the Soviet Union began to fly interplanetary spacecraft outfitted with instrumentation to detect the solar wind. The *Mariner 2* spacecraft, which the United States sent to the planet Venus in 1962, detected the wind. It found that the density of protons was about 5 per cubic centimeter. Today we are sure that the wind is electrically neutral, so the density of electrons must be the same as the density of protons. The wind speed varied from 319 to 771 kilometers per second, with an average speed of roughly 400 kilometers per second. Thus the low-density, high-speed solar wind that cometary evidence had led Biermann to hypothesize was verified.

Models of Interactions of the Solar Wind with Plasma Tails

In 1957 Hannes Alfvén (Figure 4.4) introduced magnetic fields as an important element of cometary physics. Clearly the presence of a magnetic field would increase the coupling between the cometary and solar-wind plasmas. Alfvén also introduced his idea of turning *tail rays* via the capture of magnetic field lines, a basic picture that has survived to this day (Figure 4.5). According to this view, the plasma is usually attached to the head region of the comet by the captured magnetic field lines.

Normally, the molecules flowing out from the coma of a comet within about 1.5 to 2 astronomical units of the sun are

Figure 4.4 Hannes Alfvén.

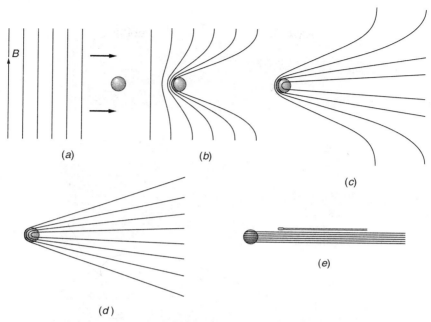

Figure 4.5 The basic interaction of the solar wind's magnetic field *(B)* with an ionized cometary atmophere. Parts *a* through *d* show the capture and wrapping of magnetic field lines by the comet, viewed perpendicular to the magnetic field. Part *e* is the view parallel to the solar-wind field; above the captured field lines, part of the field is slipping around the comet. (After Hannes Alfvén, "On the Theory of Comet Tails," *Tellus* 9 [1957]:92–96)

significantly ionized. These newly ionized molecules are trapped in the interplanetary magnetic field, which is being carried outward by the solar wind, a fully ionized proton-electron gas traveling at roughly 400 kilometers per second. The process of "picking up" these new ions slows the solar wind in the vicinity of the comet. Because the solar wind is not slowed away from the comet, the field lines wrap around the comet like a folding umbrella to form a hairpin-shaped magnetic structure with two lobes of opposite magnetic polarity. The region between the two lobes must have a current sheet to stabilize the situation. The folding of tail rays to the plasma structures, including the visible plasma tail, can be observed directly because trapped molecular ions (usually ionized carbon monoxide) serve as tracers of the field lines (see Figure 4.6).

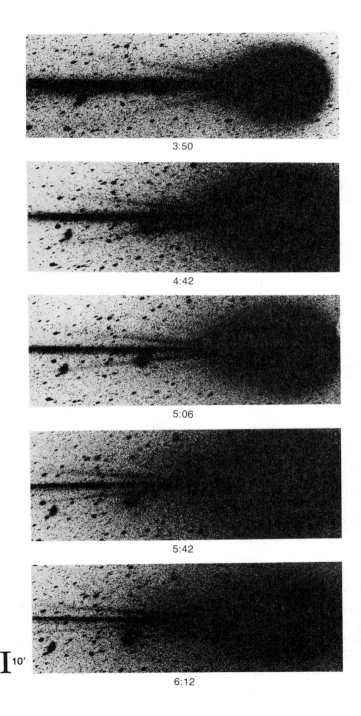

3:50

4:42

5:06

5:42

6:12

Most of the time, the field lines are rooted in the ionized gas close to the nucleus, and thus the plasma tail is attached to the comet. When the attachment is disrupted, the entire tail can become detached. This situation is called a *disconnection event* (or DE); see the discussion beginning on page 85.

Finally, comets have shock waves on the sunward side because the comet and its ionosphere are obstacles to the supersonic solar wind. These shocks lower the flow speeds and permit the solar wind to flow smoothly around the comet.

Comets as Solar-Wind Probes

The orientations of plasma tails were studied extensively by Cuno Hoffmeister through the years. He found that these tails did not point precisely away from the sun, but lagged this direction by a few degrees, in a direction opposite the comet's orbital motion. Hoffmeister also found that the angle of lag increased with the comet's orbital speed perpendicular to the radius vector. Biermann believed that the lag angles occurred because the tail acted like a wind sock as the comet's orbital motion carried it through an expanding, resisting medium—that is, the solar wind. Thus the tail orientation matches the direction of the solar wind as seen by a hypothetical observer riding on the comet. The comet's motion causes a change in the direction (called an *aberration*) just as an observer's motion seems to change the direction of raindrops falling vertically. The physics of the aberration effect is shown in Figure 4.7. This interpretation has stood the test of time and often forms the basis for using comets as solar-wind probes.

Comets indicate solar-wind or interplanetary conditions in several ways: (1) the orientations and shapes of comet tails yield solar-wind speeds through the wind-sock effect just described;

Figure 4.6 Comet Kobayashi-Berger-Milon on July 31, 1975. These photographs show a pair of tail rays closing onto the tail axis, as Alfvén's model suggests (see Figure 4.5). (Photographs by E. P. Moore and K. Jockers, Joint Observatory for Cometary Research [JOCR], operated by the Laboratory for Astronomy and Solar Physics, Goddard Space Flight Center and New Mexico Institute of Mining and Technology)

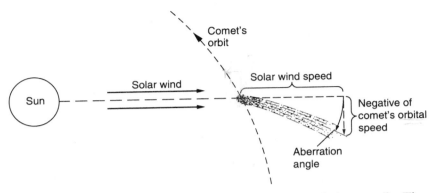

Figure 4.7 The aberration effect and the orientation of plasma tails. The comet's motion causes the tail to lag behind the prolonged line from the sun to the comet. Measurements of the aberration angle can be used to calculate the speed of the solar wind.

(2) changes in brightness can be caused by shocks encountered by the comet or by magnetic sector-boundary crossings associated with disconnection events, described below; and (3) disconnection events, if understood, could be indicators of specific conditions.

A global determination of solar-wind speeds results from an analysis of the measurements of the orientations of hundreds of comet tails. For a specific solar-wind model—that is, a radial speed, an azimuthal speed, and a polar (or meridional) speed—a set of predictions of tail orientations can be generated. A comparison of the predicted and observed orientations can yield a best model when least-squares techniques are used to minimize the differences. A global model based on 809 observations is shown in Table 4.1. The sample of observations covers the solar latitudes to plus or minus 40 to 50 degrees and heliocentric distances of approximately 0.6 to 1.4 astronomical units.

The wind-sock idea can also be applied to such specific events as a distorted tail; consider the "big bend" in comet Kohoutek shown in Figure 4.8. A detailed comparison revealed that the specific solar-wind feature responsible was a change of 30 kilometers per second in polar speed. This "big bend" in the plasma tail of comet Kohoutek was probably the first to be convincingly associated with a specific feature of the solar wind.

Table 4.1 A Solar Wind Model Derived from Plasma Tail Orientations

Parameter	Derived value
Radial speed, w_r (km/s)	400 ± 11
Azimuthal speed,[a] w_ϕ (km/s)	6.7 ± 1.7
Meridional speed,[b] w_m (km/s)	2.3 ± 1.1
RMS dispersion (degrees)	3.736

[a]The value is the equatorial value; the azimuthal speed is smaller away from the equator.
[b]The meridional speed is assumed to vary with twice the sine of the latitude and to have the same flow toward or away from the equator in both hemispheres. The value is the maximum (at 45-degree latitude) and indicates a small flow toward the equator.

Large-Scale Changes and Disconnection Events

Rapid large-scale changes involving the entire plasma tail occur in comets. These changes are uncorrelated with distance from the sun. Figure 4.9 shows a change in the orientation of the plasma tail of comet Bradfield (1979 X). The tail has

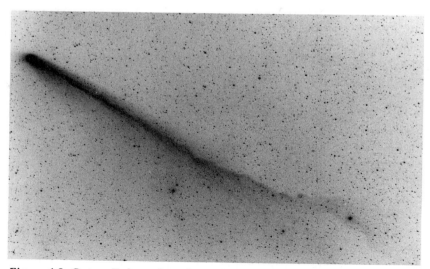

Figure 4.8 Comet Kohoutek on January 20, 1974. A "big bend" is visible in the plasma tail. (JOCR photograph)

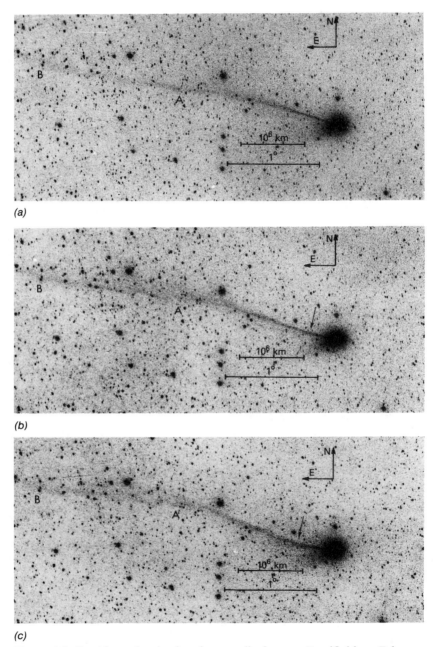

Figure 4.9 Rapid turning in the plasma tail of comet Bradfield on February
6, 1980. The times of midexposure are (a) $2^h32^m30^s$, (b) $2^h48^m00^s$,
and (c) $3^h00^m00^s$. The rapidly turning segment is marked by the
arrow. (JOCR photographs)

changed orientation by 10 degrees in 27.5 minutes, for a turning rate of about 22 degrees per hour. This event probably occurred when a solar-flare-generated change in the direction of the solar wind passed over the comet's tail.

Another fascinating aspect of the temporal behavior of comets is the disconnection event. The entire plasma tail routinely drops off a comet, drifts away, and forms again on a time scale of roughly one day; one such event is shown in Figure 4.10.

Disconnection events were accurately described by Edward Emerson Barnard early in this century. Barnard knew that there were changes in the appearance of the plasma structure, including a characteristic narrowing of the tail just before it separates from the head. Curiously, the phenomenon then was largely forgotten until Malcolm Niedner and John Brandt rediscovered it in 1978.

Disconnection events are not rare; at least one was observed in comet Halley in 1910. The 1985–1986 apparition of Halley's comet was rich in them, as we had hoped. Part of the rationale for establishing the Large-Scale Phenomena Network of the International Halley Watch was the desire to document the dramatic disconnection phenomena. During the seven months when Halley's comet was visible, some 20 obvious disconnection events and approximately 10 minor ones were recorded. (Examples are shown in Figures 6.12, 6.13, and 6.14.)

As we saw earlier (Figure 4.5), magnetic field lines are captured from the solar wind and draped around a comet's head region. The plasma tail becomes attached to the rest of the comet via the field lines that thread the ionized region of the coma. The disconnection event could be caused by three kinds of mechanisms:

1. *Ion production effects.* If the production rate were to drop drastically, the ionosphere, which holds the field lines, could disappear, allowing the field lines to slide away from the comet.

2. *Pressure effects.* A major increase in pressure on the ionosphere (caused by conditions in the solar wind) could compress the ionosphere until it was small enough to allow the field lines to slip away; or if the pressure was high enough, it could simply push the ionosphere, field lines and all, away from the head.

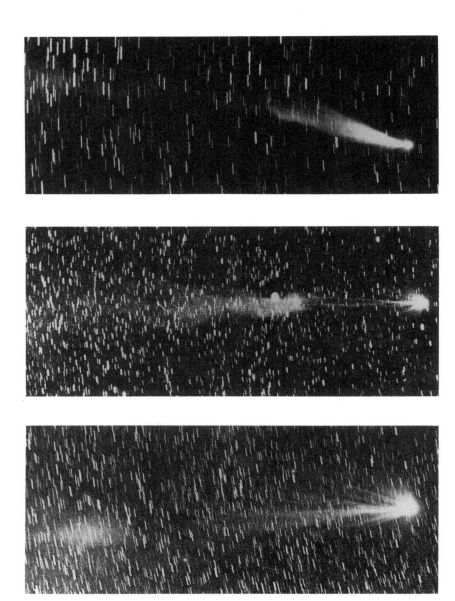

Figure 4.10 A dramatic disconnection event in comet Morehouse on (top to bottom) September 30, October 1, and October 2, 1908. The sequence shows the rejection and drifting away of the old tail and the partial formation of a new one. (Yerkes Observatory photograph)

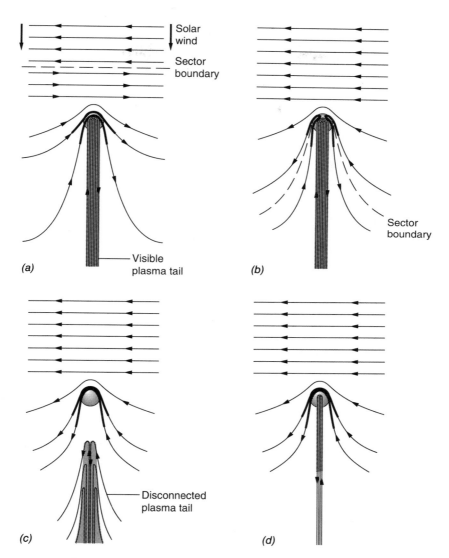

Figure 4.11 The sunward magnetic reconnection model of disconnection events: *(a)* The sector boundary separating regions of opposite magnetic polarity in the solar wind approaches a comet. *(b)* The arrival of the boundary presses a magnetic field of opposite polarity into the comet, causing the severing of field lines by reconnection. *(c)* The completely severed old tail drifts away while the comet grows a new tail. *(d)* The old tail is gone and development of the new tail continues. This sequence is repeated with each new boundary crossing. (M. B. Niedner, NASA-Goddard Space Flight Center, and J. C. Brandt, University of Colorado)

3. *Magnetic reconnection.* The magnetic field lines may be severed and then reconnected. Two logical locations for reconnection are on the sunward and tailward sides. Reconnection of magnetic field lines in either of these two locations could disconnect the tail.

Interpretation of the data is under way, but progress is complicated for a variety of reasons. We do not know if all disconnection events have a common physical cause; we do not have accurate knowledge about the solar wind or other conditions that produce these phenomena; and we do not understand the physical processes responsible for them. Nevertheless, some progress has been made.

So far, the evidence seems to favor the sunward reconnection mechanism. The obvious disconnection events tend to occur near sector boundaries that could produce the sunward reconnection. But the best discriminator may be the disconnection event of March 8, 1986, which occurred while various spacecraft were in the vicinity of the comet (as we shall see in Chapter 6). If we use this event to distinguish between disconnection mechanisms, we implicitly assume that all disconnection events have a common cause. If the sunward reconnection is the correct mechanism, a sector boundary should have passed over the comet just before March 8, and the magnetic polarity of Halley's magnetosphere should have reversed as a result of the disconnection process. The sector boundary was observed by *Vega 1* at the proper time, and the polarity reversal was recorded by *Vega 1* and *Vega 2*. Clearly, sector boundaries created by magnetic reconnection can produce disconnection events, and this model has passed a serious test. The model is shown in Figure 4.11. We cannot be sure, however, that this explanation applies to all disconnection events. A detailed investigation of all observed disconnection events is needed, and debate over the next few years is to be expected.

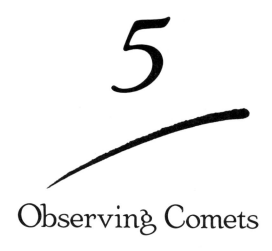

Observing Comets

The Comet Hunters

Many of the comets discovered every year are found as a result of the painstaking search efforts of dedicated amateur astronomers who look for comets just for the fun of it. We have already mentioned the late Japanese amateur Minoru Honda, who discovered or codiscovered more than a dozen comets beginning in the 1930s. This is certainly one area where amateurs have contributed as much as professionals.

Comet hunting does not require fancy equipment. The late Leslie Peltier, an amateur astronomer and successful comet hunter from Ohio, suggested that one use a telescope with a field of 1.5 to 2 degrees. Peltier outfitted his telescope with a turret eyepiece that he set up so that when different magnifications were rotated into play, they were immediately in focus. He could then quickly switch to increased magnification to examine an object of interest in more detail.

If you scan through the list of comet names in a publication such as Marsden's *Catalogue of Cometary Orbits,* you note numerous Japanese names: Mrkos-Honda, Ikeya-Seki, Tago-Honda-Yamamoto, Fujikawa, Tago-Sato-Kosaka, Suzuki-Sato-Seki, Kojima, and many others. Japan has a tradition of comet hunting that was probably started by Kaoro Ikeya, who searched the skies night after night with a homemade telescope, never becoming discouraged by a lack of success. Ikeya finally discovered his first comet after more than a year of fruitless searching and achieved well-deserved recognition. If you are interested in searching for comets, all you really need is a reasonably good pair of binoculars. Since comets tend to be brightest when they are nearest the sun, a reasonable strategy is to search near the western horizon after sunset and near the eastern horizon just before sunrise. The important thing to keep in mind is that it is possible to search for years without success. It is tempting to give up, but the successful comet hunters are those who persevere.

Anyone who does find a comet should do as Kohoutek and others have done and send a letter—or better, a cable—to the Central Bureau for Astronomical Telegrams. Staff members there will then do several things. First, they will send out a coded telegram announcing the discovery and asking other observers to verify it. Confirmed discoveries are announced through International Astronomical Union circulars (Figure 5.1). For this reason, it is essential for the object's position to be very carefully determined. If the discovery is verified, additional positional observations will be obtained and used to calculate the parameters that describe the size, shape, and orientation in space of the comet's orbit. These parameters will permit astronomers to predict future positions of the comet, using programs like the one in Appendix B. Professionals such as Brian Marsden, director of the Central Bureau for Astronomical Telegrams, devote considerable time to the process of keeping track of comets. As we will discuss later, they use very sophisticated computation methods to aid them in their work.

Sky & Telescope, an excellent magazine for amateur astronomers, has an article in its October 1987 issue titled "What to Do If You Discover a Comet" (reproduced in Appendix C). The author gives the Central Bureau's address and tells the comet discoverer how to code a telegram. Anyone beginning

Circular No. 3737

Central Bureau for Astronomical Telegrams
INTERNATIONAL ASTRONOMICAL UNION

Postal Address: Central Bureau for Astronomical Telegrams
Smithsonian Astrophysical Observatory, Cambridge, MA 02138, U.S.A.

TWX 710–320–6842 ASTROGRAM CAM   Telephone 617–864–5758

PERIODIC COMET HALLEY (1982i)

D. C. Jewitt, G. E. Danielson, J. E. Gunn, J. A. Westphal, D. P. Schneider, A. Dressler, M. Schmidt and B. A. Zimmerman report that this comet has been recovered using the Space Telescope Wide–Field Planetary Camera Investigation Definition Team charge–coupled device placed at the prime focus of the 5.1–m telescope at Palomar Observatory. Five exposures of 480–s effective duration each (in seeing measured to be 1″.0 fwhm) were taken on Oct. 16 through a broad–band filter centered at 500 nm. Definite images near the expected position and having the expected motion of P/Halley were noted. No coma was detected, and the object had a Thuan–Gunn magnitude of [g] = 24.3 $\pm$ 0.2 (corresponding to V ~ 24.2; and presumably B ~ 25). Two exposures were also made in the [r] band. Preliminary representative positions, which have an estimated external error of $\pm$ 0^s.35 in α and $\pm$ 5″ in δ but greater internal consistency, follow:

1982 UT	α_{1950}	δ_{1950}
Oct. 16.47569	7^{h}11^{m}01^{s}9	+ 9°33′03″
16.49097	7 11 01.8	+ 9 33 02
16.52153	7 11 01.7	+ 9 33 00

The object is located some 0^s.6 west of the position predicted by D. K. Yeomans (1981, <u>The Comet Halley Handbook</u>), suggesting that T = 1986 Feb. 9.3 UT. Confusion with a minor planet would be extremely unlikely. An attempt to confirm the recovery on Oct. 19 was successful in the sense that no objects were detected at the Oct. 16 locations and that the comet's image would then have been in the glare of a star; the dense stellar field has in fact thwarted other attempts to recover the comet during the past month. The recovery brightness indicates that the 1981 Dec. 18 attempt (cf. IAUC 3688) failed to record the comet by a very small margin and for an assumed geometric albedo of 0.5 leads to a radius of 1.4 $\pm$ 0.2 km. The comet's heliocentric and geocentric distances at recovery were 11.04 and 10.93 AU, respectively.

NOVA SAGITTARII 1982

<u>Corrigendum</u>. On IAUC 3736, line 17, the first astrometric position should be attributed to J. Hers, Sedgefield.

1982 October 21   Brian G. Marsden

Figure 5.1 The International Astronomical Union circular announcing the recovery of Halley's comet on October 16, 1982. (Courtesy of Harvard-Smithsonian Astrophysical Observatory)

to hunt seriously for comets should read the article. After a discovery is made, speed is crucial if the discoverer doesn't want to be scooped.

The history of comet hunting is dominated by a few observers. Mrkos and Honda, for instance, each participated in the discovery of a dozen comets. An equally successful comet hunter was Charles Perrine, who worked at Lick Observatory near San Jose, California. Perrine is most famous, however, for the part he played in an incredible coincidence. He found a comet one night in 1896 and continued to observe it for several days. Soon he received a telegram from a colleague in Germany reporting the position of the comet. A transmission error, however, caused the telegram to report a slightly erroneous position. The error was too small for Perrine to notice, so he set up his telescope to observe the comet. Incredibly, there was a comet at the erroneous position! Perrine did not realize at first that there had been a mistake. Later, when the error was revealed, Perrine found himself to be the discoverer of a second comet. Serendipity once again played a role in the progress of science.

Many amateur astronomers know the name of another comet hunter, not for the comets he discovered (he discovered 13) but for an important by-product. Charles Messier (1730–1817) kept seeing fuzzy splotches of light in the sky as he searched for comets. Again and again he had to observe them on several nights to see if they were moving or were fixed among the stars. Finally he decided to make a catalogue of the fixed objects so that he and other comet hunters would not mistake them for comets in the future. His catalogue, published in the late 18th century, listed more than 100 objects, which we now recognize as the most spectacular star clusters, nebulas, and galaxies in the northern skies. Even today, more than two centuries later, we refer to these celestial marvels by their "M numbers"—their numbers in Messier's catalogue.

Jean-Louis Pons (1761–1831) holds the incredible record of discovering or codiscovering roughly three dozen comets. Anyone who discovers that many comets is clearly an individual of great patience. Yet Pons did not exercise comparable patience in his position measurements. In 1818 he was the only observer to see a comet in the constellation Cetus. Unfortunately, his positional observations were not precise enough to permit astronomers to calculate the comet's orbital parameters, and it was lost. In 1873, however, two observers—Friedrich Winnecke in Germany and J. Coggia in France—discovered a comet almost simultaneously. Their observations were precise enough to permit good orbital parameters to be calculated. Because the comet was visible for only two weeks, however, the orbital period was uncertain. The comet's path was suspiciously close to that of the comet Pons had seen in 1818. In 1928 A. Forbes, in South Africa, discovered a comet and observed it long enough that the data led to good orbital parameters, and the period was found to be 28 years. It is now clear that all the observers saw the same comet, which is sometimes called comet Pons-Coggia-Winnecke-Forbes, though it is also often referred to as comet Crommelin, for the celestial mechanician who calculated its orbital parameters.

The astronomer Edward Emerson Barnard (Figure 5.2) was able to turn comet hunting into a profitable enterprise. H. H. Warner, a contemporary of Barnard's, offered a $200 prize to anyone who discovered a new comet. Between 1881

Figure 5.2 Edward Emerson Barnard. (Yerkes Observatory)

and 1889 Barnard claimed the prize at least 11 times and was able to buy a house with the winnings. His home became known locally as "Comet House." Incidentally, Barnard earned his living in the photography field during his years in Nashville. His success as a comet observer turned his interest more and more toward astronomy. By the turn of the century, Barnard had established himself as a first-class observational astronomer. He was one of the pioneers in astronomical photography.

Women, too, have left their mark in the annals of comet hunting. Two very successful women in the field were Caroline Herschel (1750–1848) and Maria Mitchell (1818–1889). Caroline Herschel and her brother William were brought up in a musical and intellectual household. Their father was a musician in a military band in Hanover, Germany. William emigrated to England in 1760 and began to make his way as a musician, composer, and teacher. After several years, William was joined by his sister, who served as his music assistant.

Around 1772 William began to build telescopes in his spare time, and with Caroline's assistance he started observing the skies. William's career as an astronomer overtook his career as a musician in 1781, when he discovered the planet Uranus— the first new planet to be found since the dawn of history. William was eventually appointed astronomer to the king, and Caroline continued as his assistant. Attitudes being what they were, no woman in the 18th century could be appointed to a position as prestigious as William's, but Caroline Herschel was a proficient musician and astronomer in her own right.

As it turns out, the telescopes that William Herschel built were superior even to those at the Royal Greenwich Observatory. William used his magnificent instruments to sweep the skies, cataloguing many thousands of nebulas, star clusters, and galaxies. Caroline independently used the same instruments to search for comets. On August 1, 1786, she discovered comet 1786 II, which turned out to be the first of many discoveries. In 1823 Caroline Herschel was awarded the Gold Medal of the Royal Astronomical Society for her dedicated work as a comet hunter. This is one of the world's most prestigious awards in the field. Incidentally, one of Caroline Herschel's discoveries, comet 1788 II—also known as comet Herschel-Rigollet—is a record holder. When it was observed again on its next pass near the sun and earth as comet 1939 VI, it became the comet with the longest period—154 years—to be observed on more than one pass.

Maria Mitchell is considered to be the first woman astronomer in the United States. She was born in 1818 on Nantucket Island, off the coast of Massachusetts. The islanders made their living as whalers, and Mitchell's father maintained a small observatory where he set chronometers used in navigation on long sea voyages. Mitchell came to the attention of the astronomical community when she discovered comet 1847 VI. Soon after the discovery, Mitchell became a member of the faculty at Vassar College, where she remained until her death.

These people are just a few of the luminaries in the history of comet hunting. Clearly not all successful comet hunters are highly trained professionals. Many are just stargazers who are willing to make the effort to search and search in hope of reaping the personal reward of discovering a new comet.

When the comet scientist carries out research to further understanding of comets, the effort requires the continued gathering of observational and experimental data. Observational data consist of information obtained at long distance by studying light from a comet, while experimental data consist of information obtained by instruments sent to a comet. The way data are gathered and interpreted is an important part of the story of comets.

Tools of the Comet Scientist

Fundamentally, observing techniques fall into three major classes: *spectroscopy, photometry,* and *imaging.* These techniques are practiced from both ground-based and space-based observatories and are applied to many comets. So far direct experimental techniques have been applied to only two comets, as we shall see.

Excellent sites for ground-based observatories are becoming more and more difficult to find. Many of the well-known observatories—such as Palomar, with its great 200-inch reflecting telescope—are feeling the increasing encroachment of civilization. Many cities in southern California lie near the foot of Mount Palomar. As more and more people flock to the warm

Figure 5.3 The Kitt Peak National Observatory. (National Optical Astronomy Observatories [NOAO] photograph)

California desert, the light pollution and air pollution of nearby towns have increasingly adverse effects on the skies above the observatory. Kitt Peak National Observatory (Figure 5.3), near Tucson, Arizona, has the same problem.

Fortunately, there are still a few excellent observatory sites in the world. Cerro Tololo Interamerican Observatory—administratively a part of the National Optical Astronomical Observatories—is situated in the high Andes of Chile. Astronomers from institutions all over the United States travel to Chile to carry out their research. The observing conditions also remain outstanding in Hawaii, and several observatories have been established on Mauna Kea, on the island of Hawaii.

Information about the chemical and physical conditions of a celestial body—be it a star, a planet, or a comet—is coded in the light it emits or reflects. This information is needed for an understanding of a celestial body's structure, origin, and evolution—things scientists want to know. In principle, the spectrum of a comet reveals the chemical composition of the object emitting the light, the electric and magnetic fields of the region, the object's pressures and temperatures, and the motion of the object toward or away from the observer. It's no wonder, then, that the rise of astrophysics in the 20th century was in large part spurred by advances in the study of atoms and molecules and by the development of means to observe them.

Atoms, Molecules, and Chemistry

The atom is the basic unit of ordinary chemistry. If we could magnify an atom many times, it would appear something like a miniature solar system, with a positively charged nucleus orbited by negatively charged electrons. At higher magnification still we would find the nucleus to be made up of protons, which have a positive charge, and neutrons, which have no charge. (Nuclear physics has discovered a myriad of subatomic particles, but they are not known to be important in comets.) The simplest atom, ordinary hydrogen (H), consists of a nucleus containing one proton, orbited by one electron. The deuterium atom (D), or heavy hydrogen, consists of a nucleus containing one proton and one neutron, orbited by one electron. Atoms are sometimes characterized by their atomic number and atomic weight. The *atomic number* is the number of orbiting electrons in the neutral atom, and the *atomic weight* is the number of protons and neutrons in the nucleus. We hasten to add that the picture of an atom as a miniature solar system is only an approximation. The more precise model of atoms provided by the modern quantum theory of atomic structure is not so easily visualized.

The periodic table of elements is structured by the way electrons are added to the shells in more complex atoms. After hydrogen comes helium, the next most complex atom. In undisturbed helium, the two electrons fill the innermost orbit, called the K shell, which can hold only two electrons. Hydrogen and helium make up the first row of the periodic table. The eight elements in the second row of the periodic table— lithium, beryllium, boron, carbon, nitrogen, oxygen, fluorine, and neon—add electrons to the next shell, the L shell, which can hold eight electrons. With sodium, we begin to fill the third or M shell of the atom. The third row of the periodic table consists of the eight elements beginning with sodium and ending with argon. Now things become more complicated. The M shell can hold 18 electrons, but the next element has its last electron added to the N shell. Further electrons are added to shells in a pattern that is understandable on the basis of atomic theory but is beyond the scope of our discussion.

Each column in the periodic table contains elements that are chemically similar because their atomic shell structures are similar. The elements in the first column—lithium, sodium, potassium, and so on—all contain one electron outside a closed shell and are chemically similar. The elements in the last column—helium, neon, argon, and so on—all consist of a set of filled shells and are chemically inert.

The iron atom usually has a nucleus consisting of 26 protons and 40 neutrons surrounded by 26 electrons. We say "usually" because some atoms of the same element have a different number of neutrons in the nucleus; these are *isotopes* of the same element. Hydrogen and deuterium atoms are isotopes. The chemical properties characteristic of their atomic number are quite similar, but the atomic weights (the sum of the protons and neutrons in the nucleus) differ.

The arrangement of the electrons in the atom creates an attractive force that causes atoms to combine to form molecules, such as water, H_2O. The structure of the water molecule is seen in Figure 5.4a. The arrangement is often shown schematically with short lines to represent the chemical bond, as in Figure 5.4b. Here the hydrogen is shown with one chemical bond, whereas oxygen has two. The electrons in atoms have a preference for completed shells, as exemplified by the "noble gases" helium, neon, argon, and so on. They are called "noble" because their shells are complete and they do not readily combine with other atoms to form molecules. Atoms can share electrons to form completed shells and thus produce molecules. Oxygen needs two electrons to complete a shell, and each hydrogen atom can supply one. Thus H_2O is a natural com-

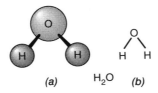

(a) H_2O (b)

Figure 5.4 The water molecule. (a) The bars represent the chemical bonds that hold the atoms of oxygen (O) and hydrogen (H) to form the water molecule (H_2O). (b) The lines represent the chemical bonds.

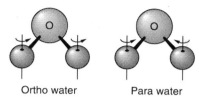

Ortho water   Para water

Figure 5.5 A water molecule takes one of two forms, depending on whether the nuclei of the hydrogen atoms are rotating in the same (ortho) direction or in opposite (para) directions.

bination of H and O. This explanation of molecules is greatly simplified. When more atoms are in the molecule or when more complex atoms are involved (carbon, for example, has four bonds), the situation is more complex.

Water molecules have another property of interest in the study of comets. The arrangement of the atoms shown in Figure 5.4 doesn't always tell the whole story, because the nucleus of the hydrogen atom can rotate, or spin. As we see in Figure 5.5, the nuclei of the two hydrogen atoms can spin in the same direction or in opposite directions. When they spin in the same direction, they produce a molecule of *ortho water;* when they spin in opposite directions, the molecule is called *para water.* These molecules produce slightly different spectra and so are observable as separate molecular species. Individual atoms change between the ortho and para forms of water very slowly.

In laboratory situations and in the nuclei of comets, the normal combination of H and O is water, H_2O. In the low-density situations found outside the nuclei in comets, encounters between atoms are rare, and the process of water formation can be incomplete; the OH, or hydroxyl, radical can be formed. Radicals are chemical combinations with unfilled bonds; they are not frequent in ordinary matter, but they are common in comets.

Another chemical structure of possible importance in comets is the polymer, a repeating arrangement or chain of simpler molecules. The molecule formaldehyde, H_2CO, can be rearranged to form a polymer. Monomers, dimers, trimers, tetramers, and so on are molecules that contains one, two, three, four, or more smaller molecular units.

Some organic molecules, generally molecules containing carbon, are considered to be prebiotic; that is, precursors to biological molecules. Carbon usually has four chemical bonds and is conducive to the production of chains and other complex structures. Complex molecules containing carbon, hydrogen, nitrogen, and oxygen can be called prebiotic because reasonable scenarios beginning with such molecules lead to the formation of molecules associated with the origin of life, such as amino acids. The general formula for an amino acid is $NH_2(CH)(R)COOH$; R (the molecular side chain) changes to give the various amino acids. The simplest amino acid is glycine, with $R = H$. Proteins are polymers of amino acids. The simple joining together of amino acids forms a peptide chain. Three coiled peptide chains twisted around each other constitute a protein molecule. In short, if we start with a simple carbon molecule and an elementary knowledge of bonding, we can proceed logically to the formation of complex biological molecules.

The Nature of Light

A great deal of what we can know about comets comes from an analysis of their light. It is useful, therefore, to review a little of what we know about the nature of light.

Light remains a fascinating mystery. It behaves sometimes as a wave and sometimes as a particle. When light acts as if it were a wave, it behaves as a transverse wave (Figure 5.6). If you shake a rope, a wave appears to move down the length of it. If you look carefully at what is happening, though, the material of the rope is moving perpendicular to the rope's length. The wave moving along the rope is just a disturbance. The same thing happens with water waves. A surfer out in the ocean beyond the point where waves are breaking will just bob up and down, so the motion of the surfboard is up and down. A breaking wave is no longer a transverse wave. The wave is modified when the depth of the water becomes less than the length of the wave; this is when the surfer can have fun.

The actual disturbance of the light wave is electromagnetic. Colors that we see depend on the wavelengths of the light. The wavelength of visible light, the distance between "crests" (or "troughs") in the wave, is tiny. For yellow light the wavelength is about 0.000023 inch. The wavelength of red light

Plate 1 Halley's comet as photographed on March 7, 1986. This image also captures the moon just above the horizon. *(Courtesy of Greg Polus, Deer, Arkansas)*

Plate 2 Halley's comet as photographed on March 20, 1986. The extraordinary amateur images in Plates 1 and 2 capture the appearance of the comet to the general observer. *(Courtesy of Greg Polus, Deer, Arkansas)*

MALUM ASTRUM HAROLDO PRAEFIGURATUM A.D.1066 = PRO STELLAE RESURRECTO A.D.1986 DELPHINE DELSEMME FECIT

Plate 3 Facsimile reproduction of the part of the Bayeux tapestry showing Halley's comet. *(Courtesy of Delphine Delsemme)*

Plate 4 The Giotto fresco, *The Adoration of the Magi*, in the Scrovegni Chapel, Padua, Italy. Because Halley's comet had appeared in 1301 and the fresco was completed probably in 1304, the comet may have been the model for the star of Bethlehem. *(Courtesy of Scala/Art Resource)*

COMET HALLEY

Pioneer Venus Orbiter
2-6 February 1986

Plate 5 Color rendering of the hydrogren cloud around Halley's comet based on scans of Lyman-alpha intensity obtained from the *Pioneer Venus Orbiter* from February 2 to 6, 1986. The immense size of this cloud is shown by the filled circle at bottom left, which has the same size as the disk of the sun at the comet's distance. *(Courtesy of A. I. F. Stewart, Laboratory for Atmospheric and Space Physics, University of Colorado)*

Plate 6 Comet West as photographed on March 7, 1976, from New England. The dust tail (left) appears white, while the plasma tail (right) appears blue. *(Courtesy of Betty Milon)*

Plate 7 Comet Halley as photographed on March 8, 1986, from Easter Island, as part of the International Halley Watch. The dust tail (above) appears white, while the plasma tail (below) appears blue. *(Courtesy of William Liller, Large-Scale Phenomena Network, International Halley Watch)*

Plate 8 Comet Halley as photographed on April 14, 1986, from Easter Island, as part of the International Halley Watch. The plasma tail points directly upward, and the dust tail fans out to the left. The image also captures the globular cluster Omega Centauri (bottom). *(Courtesy of William Liller, Large-Scale Phenomena Network, International Halley Watch)*

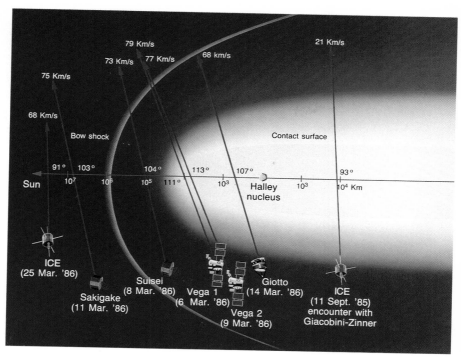

Plate 9 Summary of spacecraft to comets Halley and Giacobini-Zinner. The labels give the date of closest approach, the angle of the trajectory with respect to the sun-comet line, and the relative speed of the encounter. Some cometary features are shown schematically; note that the scale is in increments of a factor of ten. *(NASA)*

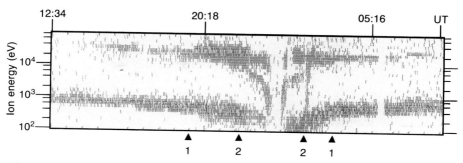

Plate 10 Plasma results at Halley's comet obtained from the ion analyzer on board *Giotto*. The times shown are on March 13 and 14, 1986; 12:34 on the 13th corresponds to 2.8 million kilometers before closest approach, and 12:00 on the 14th (end of data shown) corresponds to 2.9 million kilometers after closest approach. Blank spaces are due to data gaps. The ion energy is in electron volts (eV); the data at about 3×10^4 eV are from the "pickup ions," and the data at about 1×10^3 eV are from protons in the solar wind. The fluxes are color-coded, with red indicating the highest and dark blue the lowest. The variation of the fluxes from the cometary pickup ions and the solar-wind protons is shown throughout the encounter as viewed 50 to 60 degrees from *Giotto's* spin axis. See Chapter 6 for a detailed discussion of the data. *(Courtesy of Alan Johnstone, Mullard Space Science Laboratory, University College London)*

Plate 11 Wide-angle view of the Milky Way and comet Halley on April 8, 1986. Photograph taken as part of the Can-Do Project on board the *Kuiper Airborne Observatory* while flying over New Zealand. *(Courtesy of the C. E. Williams Middle School, South Carolina)*

(a)

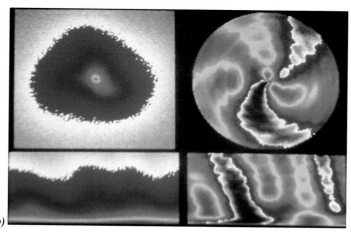

(b)

Plate 12 Cyanogen jets and modern image processing. *(a)* Image of Halley's comet on April 23, 1986, in light emitted by cyanogen (CN). North is up and east is to the left. Jets are faintly visible to the north and to the southeast; a detached structure is visible to the west. *(b)* A four-step process of enhancing the jet. First (upper left), the image in *(a)* is displayed in false color. The jets are no longer visible, and we see only a kidney-shaped coma. Second (lower left), the image is transformed to a polar coordinate system. Radius increases upward and angle increases to the right. Each horizontal row in this panel corresponds to a circle centered on the nucleus in the upper left panel. Third (lower right), each horizontal row in the previous panel has intensities scaled to run from 0 to 255. Because the intensities decrease with increasing radius, the radial variation of brightness is effectively eliminated. The result is displayed in false color. Fourth (upper right), the previous panel is transformed back to the ordinary coordinate system on the sky. The CN jets are now clearly visible. They have been interpreted as CN radicals coming directly from small dust grains. *(Image processing by D. A. Klinglesmith III, Laboratory for Astronomy and Solar Physics, NASA-Goddard Space Flight Center. Original image and illustration courtesy of M. F. A'Hearn, University of Maryland)*

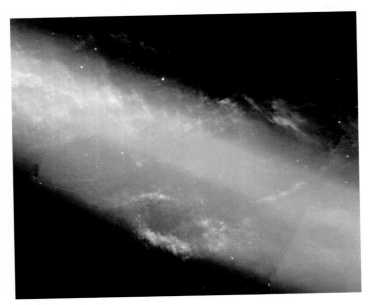

Plate 13 View of the ecliptic plane as seen by the *Infrared Astronomical Satellite (IRAS)*. The false color image is constructed from 12-micron (blue), 60-micron (green), and 100-micron (red) scans and is 30 degrees wide. The broad blue haze is the zodiacal dust, and transecting it is the central asteroidal dust band. The comet Tempel 2 dust trail is seen in the lower part of the diagram with the comet's coma at left. Debris has been detected along the orbit of comet Tempel 2 from 2.5 degrees ahead of the comet to at least 60 degrees behind it. The red, cloudlike structures are called "infrared cirrus" and originate in our Milky Way galaxy. *(Courtesy of M. V. Sykes, Steward Observatory, University of Arizona)*

Plate 14 Artist's rendering of the *Comet Rendezvous Asteroid Flyby* spacecraft at the comet. The spacecraft is near a dark, irregular surface that is emitting jets of material. *(Courtesy of NASA-Jet Propulsion Laboratory)*

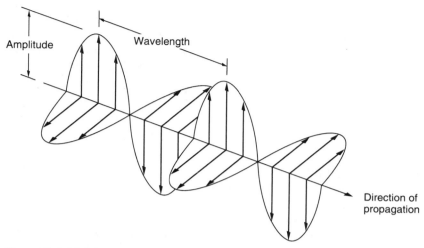

Figure 5.6 Light as a transverse wave. The variations of electric and magnetic fields—the electromagnetic disturbances—are at right angles to the direction of propagation. Here two perpendicular sinusoidal waves are shown in perspective.

is longer than the wavelength of blue light. The approximate wavelengths of colors are shown in Table 5.1. The unit of wavelength used here is the angstrom (10^{-8} centimeter).

When light is considered as a particle, the particles are called photons. The photons are distinguished by their energy, which is inversely proportional to the wavelength. Thus, while blue light has a shorter wavelength than red light (when light is considered as a wave), blue light has photons of higher energy than red light (when light is considered as a particle).

Table 5.1 Color and Wavelengths

Color	Wavelengths (angstroms)
Violet	3800–4400
Blue	4400–5000
Green	5000–5600
Yellow	5600–5900
Orange	5900–6400
Red	6400–7500

Light is but one small part of the electromagnetic spectrum; in essence, the part that can be detected with the human eye. At wavelengths longer than red light (and hence involving lower-energy photons), we have in turn the infrared and radio waves. At wavelengths shorter than violet light (and hence involving higher-energy photons), we have in turn ultraviolet light, X rays, and gamma rays. All are part of the electromagnetic spectrum, but with different wavelengths or energies.

It must be remembered that the wave and particle descriptions of light are completely complementary. The one that a scientist adopts to explain a phenomenon often depends only on convenience.

The spectroscopy of atoms and molecules is one case that is easily explained by the particle nature of light. The electrons surrounding the nucleus of an atom can have different energy levels, somewhat like different orbits around the nucleus. When an atom "absorbs" a light photon, an electron goes from a lower to a higher energy level. When an atom "emits" a light photon, an electron goes from a higher to a lower energy level. An analogous situation exists for molecules. The critical thing about these properties is that the absorption or emission of light occurs at a precise wavelength or energy that is unique to the specific atom or molecule involved. The spectrum of the sodium atom, for instance, is different from that of all other atoms, and the spectrum of water is different from that of all other molecules. These unique spectra are the basis for spectroscopic techniques used to determine the composition and physical properties of celestial bodies.

Spectroscopy

The stage was set for the science of spectroscopy when Newton discovered, through a brilliant series of experiments, that seemingly white sunlight is actually made up of the complete rainbow, or spectrum, of colors: red, yellow, green, blue, and violet (Figure 5.7). Today we know a great deal more about spectroscopy. The visible colors are only a small part of the spectrum, which also contains radio, microwave, infrared, ultraviolet, and X radiation. We have also learned that atoms and molecules absorb and emit light at very specific wave-

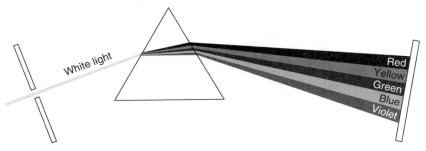

Figure 5.7 Newton's spectral experiment. Sunlight (or white light) passed through a prism is broken into a spectrum.

lengths (colors). The patterns are distinctive enough to permit us to identify the composition of a celestial object from a study of its spectrum.

A *spectrograph* is an optical device that splits light into its component wavelengths (or colors) for analysis. A simple example is the breakup of a beam of sunlight into a rainbow of colors when it passes through a glass prism. When we turn a spectrograph on a comet, what we see depends on where in the comet we are looking. The first visual observations of the spectra of comets were made in the 1860s, and the first photograph of a cometary spectrum was taken in 1881. The spectra of comets show emission lines often superimposed on a dark-line spectrum identical to the sun's.

Much research on the nature of light as it is emitted by various types of objects has shown that a hot solid or a hot, highly compressed gas emits a continuous spectrum—that is, a spectrum that contains no dark lines. If we take the light from a source of continuous radiation and pass it through a cooler vapor, that vapor will absorb just the colors that it would emit if we were to look at the vapor alone (Figure 5.8). The interior of the sun is an immensely hot, compressed gas that emits a continuous spectrum, and its hot, gaseous outer layers are the absorbing vapor that accounts for the dark lines. If we compare the pattern of dark lines in the solar spectrum with the patterns of bright lines observed in the laboratory, we can identify all the elements that make up the sun's outer layers.

The same can be said of the light emitted by other stars and of sunlight reflected by the planets and absorbed in their atmospheres. Additional information coded in the light of

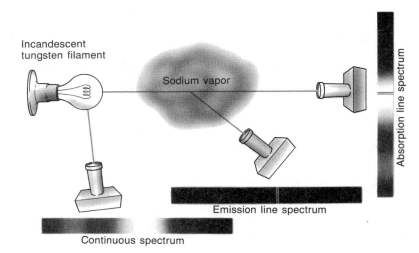

Figure 5.8 Spectrographs viewing three different physical situations (the lines in the diagrams represent the viewing directions of the spectrographs). A spectrograph viewing an incandescent solid—the tungsten filament—sees a continuous spectrum. A spectrograph viewing a cloud of glowing gas sees a spectrum consisting of emission lines characteristic of the atoms in the gas. A spectrograph viewing an incandescent solid through a cloud of glowing gas sees a continuous spectrum crossed by dark absorption lines; the wavelengths of the absorption lines are the same as the wavelengths of the emission lines characteristic of the atoms in gas. (After J. C. Brandt and S. P. Maran, *New Horizons in Astronomy* [New York: W. H. Freeman, 1972])

celestial objects tells us a lot about the situation of the atoms in the gas that emits or absorbs the spectrum lines we observe. In principle, we can infer the abundance of the elements and the temperature and density of the gas and tell whether it is moving toward or away from the observer by measuring the *Doppler effect*—the change that occurs in the frequency of the waves when the object and the observer are moving in relation to each other. (We will discuss the Doppler effect shortly.) We can also tell whether the gas is turbulent or quiescent (Doppler effect again), and whether there are magnetic or electric fields in it. In practice, these effects can be very complex, and detailed analysis is often needed to extract them.

In the observatory we use a spectrograph to produce images of the spectrum of a comet or whatever other celestial

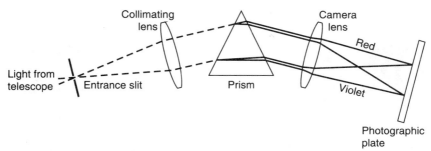

Figure 5.9 The prism spectrograph. A lens system takes light gathered by the telescope, passes it through a prism, and brings it to a focus. Because violet light is bent more than red light, for instance, as it passes through a prism, separate images of the various colors are produced on the photographic plate. The bending in the camera lens is exaggerated for clarity. (After J. C. Brandt and S. P. Maran, *New Horizons in Astronomy* [New York: W. H. Freeman, 1972])

object we are studying. We can then take the images back to the laboratory for further study. The spectrograph contains several optical components in addition to the prism used to disperse the light. In fact, most modern spectrographs use diffraction gratings instead of prisms, but the general result is the same. The spectrograph (Figure 5.9) is attached to the telescope, which collects the light of the faint comet and sends it into the spectrograph.

The spectra of the heads of comets (e.g., Figure 5.10) show emission bands characteristic of molecules such as molecular carbon (C_2, C_3), cyanogen (CN), and CH. Some comets show the emission lines of sodium. The emission lines of molecules identified in the fainter spectrum of comets' tails are all produced by ionized molecules—that is, molecules that have lost an electron. The primary emissions observed are due to ionized carbon monoxide (CO^+) and molecular nitrogen (N_2^+). The spectrum of a Type II tail shows a reflected solar spectrum, and from this fact we infer that Type II tails are made up primarily of dust particles that simply reflect (or scatter) the sunlight that falls on them. These characteristics of cometary spectra and the way they change with the comet's distance from the sun contain additional clues to the nature of comets.

Over many years of study we have built up a fairly consistent picture of cometary spectra. The molecular carbon and

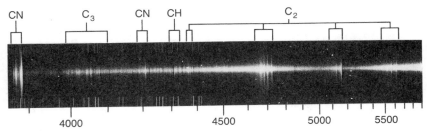

Figure 5.10 Representative spectrum of the head region of a comet (eclipse comet 1948 XI). The wavelengths are given in angstroms, and many of the observed species are marked. (McDonald Observatory spectrogram taken by P. D. Jose and P. Swings. Courtesy of C. Arpigny, Institut d'Astrophysique, Liège, Belgium)

cyanogen emitted by the head are the first spectra to become visible as the comet approaches the sun; cyanogen can be detected when the comet is as far as 3 astronomical units from the sun. When the comet reaches 1.5 astronomical units from the sun, the ionized carbon monoxide appears in the tail. Comets that approach much nearer to the sun show emission lines of sodium. If the comet gets to within 10 percent of the earth's distance to the sun, lines of iron and nickel appear in the spectrum. Table 5.2 lists the atoms and molecules observed in cometary spectra and detected by in situ measurements made by spacecraft up to the present time.

The Doppler Effect

Spectroscopy can also be used to detect the motion of an emitting body moving directly toward or away from the observer, because of the Doppler effect—a change in the wavelength of light caused by motion. If the light source is moving, the beginning and end of a wave are emitted at different locations. As Figure 5.11 shows, when the source is moving toward the observer, the wavelength is shortened, and when the source is moving away from the observer, the wavelength is lengthened. Because the original wavelength is solely a property of the emitting or absorbing atom, the shift in wavelength produced by the motion can be measured precisely. The shift in wavelength as the light source moves toward or away from the observer increases with the speed of the movement, so the speed can be determined from the spectra.

Table 5.2 Chemical Species Identified in Comets

Atoms	Molecules	Positive ions	Possible identifications
Hydrogen (H)	Diatomic carbon (C_2)	Carbon ion (C^+)	Sulfur ion (S^+)
Oxygen (O)	CH	Oxygen ion (O^+)	Diatomic sulfur ion (S_2^+)
Carbon (C)	Cyanogen (CN)	Calcium ion (Ca^+)	Diatomic oxygen ion (O_2^+)
Sulfur (S)	Carbon monoxide (CO)	Carbon monoxide ion (CO^+)	HCO^+
Sodium (Na)	Carbon dioxide (CO_2)	Carbon dioxide ion (CO_2^+)	Carbon sulfide ion (CS^+)
Potassium (K)	Carbon sulfide (CS)	Hydroxyl ion (OH^+)	Carbon disulfide ion (CS_2^+)
Calcium (C)	NH	CH^+	Sodium ion (Na^+)
Vanadium (V)	Hydroxyl (OH)	Cyanogen ion (CN^+)	Iron ion (Fe^+)
Manganese (Mn)	Triatomic carbon (C_3)	Diatomic nitrogen ion (N_2^+)	$H_5C_2O^+$
Iron (Fe)	NH_2	Water ion (H_2O^+)	$H_5C_2O_2^+$
Cobalt (Co)	Water (H_2O)	Hydrogen sulfide ion (H_2S^+)	$H_7C_3O_2^+$
Nickel (Ni)	Hydrogen cyanide (HCN)	Hydronium ion (H_3O^+)	$H_6C_3O_3^+$
	Methyl cyanide (CH_3CN)		$H_9C_4O_3^+$
	HCO		Polymerized formaldehyde ($(H_2CO)_n$)

We frequently observe the Doppler effect on sound waves. We notice that the pitch of a train whistle or car horn rises as the moving source approaches us and then drops as it races past us and recedes. This general phenomenon was explained by the Austrian physicist Christian Doppler in 1842 and was named in his honor.

Photometry

Photometry, as the word implies, is the measurement of light. We can measure the light from an entire comet or from a small part of it. The light is actually detected by a sensitive photoe-

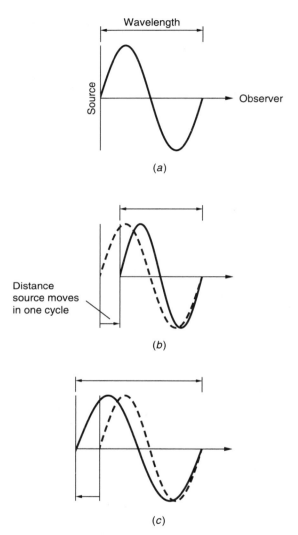

Figure 5.11 The Doppler effect. The wavelength (and frequency) of an emitting source as perceived by an observer is altered by motion with respect to the observer. *(a)* Stationary source; no alteration. *(b)* Source moving toward observer; wavelength is shortened. *(c)* Source moving away from observer: wavelength is lengthened. (After J. C. Brandt and S. P. Maran, *New Horizons in Astronomy* [New York: W. H. Freeman, 1972])

lectric device of some sort (Figure 5.12). The photoelectric device converts light into an electric current, which we then measure. To be able to calculate the intensity of light that produces a measured current, we must have calibrated our measuring device; that is, we must have measured in the laboratory the current produced by some known light source.

The comet's light, before it reaches the photometer, is usually passed through a specialized filter that transmits only that band of wavelengths of special interest to the researcher. The "filter" can be a spectrograph—that is, we can scan a photometer over the spectrum produced by a spectrograph. A plot of the current produced by the photometer is then a plot of intensity versus wavelength in the spectrum (Figure 5.13).

Imaging

The earliest images of comets were simply drawings of naked-eye views or, later, of views seen through the eyepiece of a telescope. Today photography is the primary method for producing images. Particularly in cases of very faint comets, charge-coupled device (CCD) imaging is becoming more common; see Figure 6.7 for an example.

The nature of a photograph of a comet depends on the scientific objective. If a researcher is interested in measuring the position of the comet in relation to the stars, so that the result can be put in a computer program to ascertain the parameters of the comet's orbit, then a short exposure is called for. The resulting image records the relatively small, bright part of the head surrounding the nucleus and is ideal for positional measurements.

If a researcher is interested in the faint features of a comet's tail, a long-exposure photograph is in order. Usually in such photographs the head region is considerably overexposed and the photograph is useless for research on the head. Often the researcher will want to make an image that shows the distribution of a certain emission, say the blue emission of ionized carbon monoxide. The researcher will then choose the appropriate combination of photographic emulsion and filter to isolate the emission.

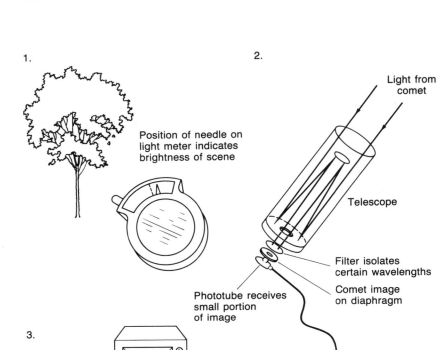

1.

2.

Light from comet

Position of needle on
light meter indicates
brightness of scene

Telescope

Filter isolates
certain wavelengths

Comet image
on diaphragm

Phototube receives
small portion
of image

3.

Comet's
brightness plotted
on moving chart

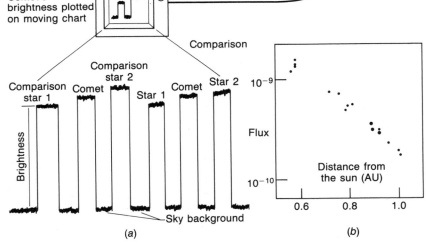

Comparison

Comparison
star 2

Comparison   Comet   Comet   Star 2 10^{-9}
   star 1   Star 1

Flux

Brightness

10^{-10}

Distance from
the sun (AU)

Sky background

0.6   0.8   1.0

(a)   (b)

Figure 5.12 Photometry. *(a)* The photoelectric process. 1. Analogy to light
meter. 2. Method of obtaining observations. 3. The raw observing record. *(b)* The end result, a graph showing the variation in
brightness of cyanogen (CN) emission from comet Kohoutek
(1973 XII) as a function of distance from the sun. (Data by
Luboš Kohoutek, Hamburg Observatory)

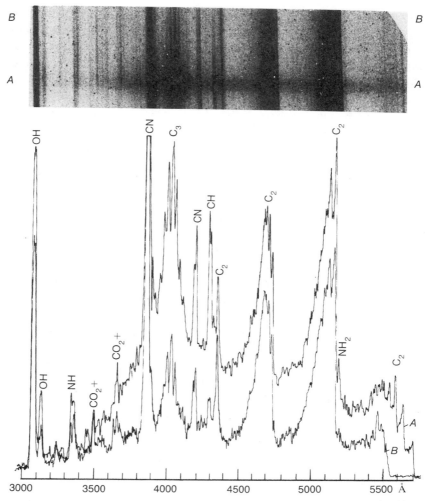

Figure 5.13 The spectrum of comet Bradfield (1979 X) on February 5, 1980. *A* is at the nucleus and *B* is 17,000 kilometers in the tail direction. (Courtesy of S. M. Larson, Lunar and Planetary Laboratory, University of Arizona)

The Study of Comets from Space

The data available from ground-based observations are crucial but limited. Many abundant elements have important spectrum lines in wavelengths that do not penetrate the earth's atmosphere. One of them is hydrogen, with its most important *(resonance)* line at 1216 angstroms (Å). The limitation imposed by the earth's atmosphere can be overcome by observations from rockets, spacecraft in earth orbit, or spacecraft in orbit around other planets. Although rockets overcome the limitations of the atmosphere, their own limitation is a maximum observing time of a few minutes. We described the discovery of the hydrogen-hydroxyl cloud on comet Tago-Sato-Kosaka by the *OAO-2* spacecraft in Chapter 3.

Comet Kohoutek and Skylab

The first of the two comets that Luboš Kohoutek discovered in March 1973 remained faint and disappeared into deep space without causing any stir. The second one was a different matter. When the early estimates of its orbit became available, it was found that the comet would come very close to the sun and reasonably close to the earth in December 1973 and January 1974. Using standard methods to estimate the brightness the comet might be expected to reach, the experts guessed that it would be the brightest object in the sky, other than the sun and moon, around Christmas 1973. If it had reached that estimated brightness, comet Kohoutek would have been one of those rare comets that is so bright that it is visible in broad daylight.

Of course, the brightness of the comet fell far short of those early estimates. Many people living in the northeastern United States never saw it. Most of us who live in that part of the country live in or near urban areas, and the air pollution and light pollution of our cities and towns simply overwhelmed the light of the not-so-bright comet. Those of us fortunate enough to spend January 1974 on a mountaintop in the southwestern United States were treated to a beautiful naked-eye view. Since most Americans don't live in the Southwest, the comet fizzled in the public relations department.

The experts who estimated the brightness of the comet were working with the best data available to them. But making such estimates is like trying to make a living on Wall Street. You do the best research you can and use the best data you can, but there is always a considerable degree of uncertainty. We try to predict the behavior of a newly discovered comet such as Kohoutek by using data from other comets, but comets are not all alike. The experts who have worked with comets for years did give due notice of the uncertainties in the brightness estimates.

Although comet Kohoutek was a disappointment from the public's point of view, it was a scientific success. Normally a new comet is discovered days or weeks before it reaches maximum brightness. Comet Kohoutek, however, was discovered nearly 10 months before maximum, so astronomers had ample time to plan a well-coordinated research program. It was fortunate that the manned *Skylab* spacecraft was in orbit at the time, so that NASA was able to reprogram some of the astronauts' activities to focus on the comet. As a result of the intense scientific activity, several new molecules—for example, methyl cyanide (CH_3CN) and hydrogen cyanide (HCN)—were identified at radio wavelengths, thus establishing radio studies as an important area of cometary research, and vital new information was obtained about the dynamics of the comet's tail. Photographs of comet Kohoutek (Figure 5.14) also revealed

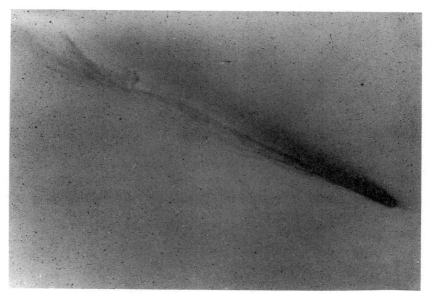

Figure 5.14 Comet Kohoutek on January 11, 1974. The "Swan cloud" (lower right) is some 0.1 astronomical unit from the head. (JOCR photograph)

interesting phenomena in the plasma tail. We will return to some of this information later.

IUE and PVO

IUE and PVO stand for *International Ultraviolet Explorer* and *Pioneer Venus Orbiter,* spacecraft that have provided the opportunity for extensive ultraviolet observations of comets. *IUE,* launched in January 1978, carries out spectroscopy in the wavelength range 1150 to 3200 angstroms. Observations are expected to be carried out until the spacecraft dies. As of 1991, plans were being made for a 14th year of operation. At least 26 comets have been observed by *IUE.* Major results include the discovery of diatomic sulfur (S_2) in comets and the general observation that cometary spectra in this ultraviolet region (being dominated by hydrogen Lyman-alpha [Lyman-α] emission at 1216 angstroms and hydroxyl [OH] bands around 3090 angstroms) tend to be rather similar.

PVO was placed into orbit around Venus in December 1978. The ultraviolet instrument can be used in a mode that allows extended mapping of the hydrogen Lyman-alpha emission. Comets Encke, Giacobini-Zinner, Halley (see Plate 5), and Wilson have been mapped and total gas emission rates determined. The useful life of *PVO* is expected to end in approximately September 1992.

Until the 20th century, new ideas about comets came slowly and old interpretations, often based on limited observations or conjecture, could persist for years or even centuries. The pace of change quickened with the introduction of photography and spectroscopy. When spacecraft were sent to observe comets, our understanding changed in a matter of days. If an active program of cometary research continues, we can expect an ever-increasing rate of change.

In Situ Measurements

Until quite recently, cometary data were limited largely to those obtained by "remote sensing" through instruments designed for use here on earth. Typical resolution of ground-based observations is approximately 1 arc second, which corresponds to a linear dimension of 300 kilometers on an object 0.4 astronomical unit from earth. Typical dimensions of the nucleus of Halley's comet, however, are 10 to 15 kilometers. Thus ground-based observations provided little hope of directly probing the source of all cometary phenomena, the nucleus. Now, however, we can send instruments directly to comets and make *in situ measurements*.

In situ observations permit direct measurement of the densities and energies of plasmas (ionized gases) by means of devices called mass spectrometers (Figure 5.15). Electric and/or magnetic fields are used to determine these quantities as a function of *M/Q;* that is, the mass *(M)* of the ion divided by its charge *(Q)*. The physical characteristics of electrons (numbers and energies, for example) also can be measured. When neu-

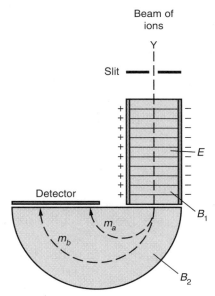

Figure 5.15 A mass spectrometer. The beam of ions passes through the slit and then through a region of perpendicular electric field E and magnetic field B_1. Ions with specific speeds (equal to E/B_1) pass into the second region, with magnetic field B_2 and no electric field. The ions spiral to the detector where the mass m_a is smaller than m_b. Mass spectrometers measure the number of ions with mass per unit charge in different ranges. Many variations are used in practice.

tral gases are to be measured, they are ionized by the instrument before they enter the mass spectrometer.

The composition of dust particles can be determined by letting them hit a metal plate. They are ionized by the impact and analyzed by mass spectrometers. Dust counters have recorded the fluxes (number of particle impacts per square centimeter per second), the momenta (mass times velocity) involved, and the distribution of different particle masses.

Magnetometers can measure both the strength and the direction of the magnetic field. Devices such as search coils and antennas can be used to measure plasma wave activity. And special techniques can be employed to measure high-energy particles in comets.

The opportunity provided by space missions to probe the nucleus and make myriad in situ measurements offered the exciting promise of a new era in cometary physics, and clearly this promise has been fulfilled. The direct exploration of comets began on September 11, 1985, when the *International Cometary Explorer (ICE)* spacecraft passed through the plasma tail of comet Giacobini-Zinner. This spacecraft was first launched on August 12, 1978, as part of a three-spacecraft mission to study the interaction of the solar wind with the earth's magnetosphere; at launch it was named the *Third International Sun-Earth Explorer (ISEE-3)*. Instruments on board the *ISEE-3* spacecraft monitored the input of solar wind to the magnetosphere from a location toward the sun, while the other two spacecraft *(ISEE-1* and *ISEE-2)* were located inside the magnetosphere, where they measured the response in the magnetosphere to changes in the solar wind. The ISEE mission spacecraft completed four years of highly successful operation in this role.

When it became apparent that the United States would not mount a dedicated mission to Halley's comet, study of the possibility of diverting *ISEE-3* to the comet was accelerated. The solution to the problem of sending *ISEE-3* to the comet was the brainchild of Robert Farquhar at NASA's Goddard Space Flight Center. Farquhar's solution involved sending *ISEE-3* into an orbit where it made five encounters with the moon, each of which added to the spacecraft's energy. The final encounter, only 120 kilometers above the lunar surface, propelled the spacecraft on a long excursion through the earth's

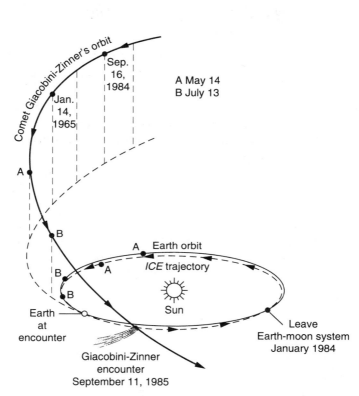

Figure 5.16 Perspective view of the *ICE* trajectory to comet Giacobini-Zinner showing the encounter circumstances. (NASA-Goddard Space Flight Center)

magnetic tail and then off to encounter comet Giacobini-Zinner. Because the spacecraft's mission had been irrevocably altered, its name was changed to the *International Cometary Explorer (ICE)*.

The intercept geometry is shown in Figure 5.16. At intercept, the *ICE* spacecraft and comet Giacobini-Zinner were 0.47 astronomical unit from earth, 1.03 astronomical units from the sun, and near the plane of the ecliptic, or the plane of the earth's orbit. The closest approach to the comet took place on September 11, 1985, at 11:02 UT (universal time, or Greenwich mean time), when the spacecraft passed at a distance of 7800 kilometers from the nucleus in the direction of the comet's tail. The tailward intercept of the comet was chosen because the instrumentation was nearly ideal for probing the environment of the plasma tail and because the other missions to comet Halley were being planned to pass on the sunward side of that comet's nucleus. As viewed from comet Giacobini-Zinner, *ICE* traversed the tail moving south to north roughly perpendicular to the ecliptic, at a relative speed of 21 kilometers per second.

The spacecraft survived the encounter with no discernible damage. The principal concern before the encounter had been that relatively high-speed collisions with cometary dust particles might degrade the spacecraft's solar cells and cause a dangerous reduction in its electrical power. After the encounter with Giacobini-Zinner, the spacecraft's trajectory took it on to a distant encounter with comet Halley, some 0.2 astronomical unit sunward, on March 25, 1986. The *ICE* spacecraft continues to move through the solar system, and its orbit is expected to bring it close to the moon on August 10, 2014.

6

In Pursuit of Comets Giacobini-Zinner and Halley

Comet Giacobini-Zinner: A Prelude to Halley

The tailward traverse of comet Giacobini-Zinner on September 11, 1985, gave us our first in situ look at the cometary environment. Many of the features observed were expected, but there were surprises, too. The comet was discovered in 1900 by Michel Giacobini at Nice, on the French Riviera, and was rediscovered in 1913 by Ernst Zinner at Bamberg, Germany. It moves in a path that takes it around the sun once every 6.5 years in the direct *(prograde)* sense. Its perihelion distance is 1.03 astronomical units, and its orbital inclination to the plane of the ecliptic is 32 degrees.

At the time of the encounter between comet Giacobini-Zinner and the *ICE* spacecraft, the rate at which gas was produced by sublimation of nuclear ices was estimated at 3×10^{28} water molecules per second. Photographs taken at earlier appearances (Figure 6.1) showed that a plasma tail and plasma phenomena were to be expected. Yet the comet's appearance in 1985 did not show the classic plasma tail (Figure 6.2).

The *ICE* observations of comet Giacobini-Zinner can be understood in terms of a model proposed by the Nobel Prize–winning physicist Hannes Alfvén. In this model, the ionized molecules produced by the comet become an obstacle in the flowing solar wind, and the interplanetary magnetic field in the solar wind is captured by the comet. The motion of the solar wind drapes the magnetic field lines around the comet to form a hairpin-shaped configuration. The result is a field that has two lobes of opposite polarities on opposite sides of the comet. Along the tail of the comet, where the two lobes meet, there is an abrupt change in magnetic polarity. If this abrupt polarity change is to be physically maintained, a current sheet must form. The *current sheet,* as the name implies, is a flat, sheetlike distribution of electrified particles flowing across the

Figure 6.1 Comet Giacobini-Zinner on October 26, 1959. The plasma tail is clearly visible. (Official U.S. Navy photograph obtained by E. Roemer, University of Arizona)

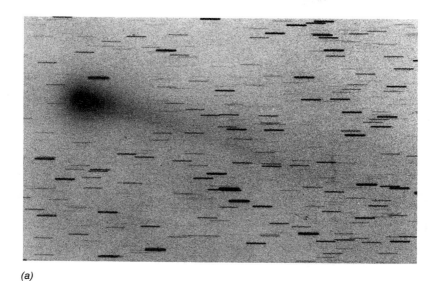

(a)

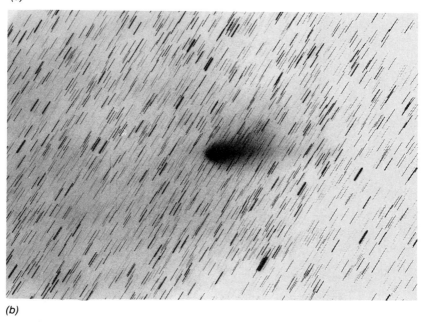

(b)

Figure 6.2 The appearance of comet Giacobini-Zinner in 1972 and 1985 did not clearly show the plasma tail. *(a)* The comet as photographed on July 13, 1972, by F. Dossin. (Courtesy of Haute-Provence Observatory/CNRS) *(b)* The comet on September 12, 1985, showing a tail over 450,000 kilometers long. (NOAO photograph by S. Majewsky, Yerkes Observatory, University of Chicago)

comet's tail. The visible manifestations of the plasma tail are caused by emissions from molecules trapped in the captured magnetic field. This is the model shown in Figure 4.5.

The in situ measurements verified the presence of extensive plasma phenomena. The magnetic field showed the draped topology and the current sheet predicted by Alfvén. Some sample field data are shown in Figure 6.3. The field orientations are just what would be expected from the model. The radial (to the sun) component of the magnetic field was negative on the inbound leg of the spacecraft's passage through the tail, just as we would have expected from the polarity of the solar-wind flow. At 11:02 UT the spacecraft was at its closest approach to the comet's nucleus, and the radial component of

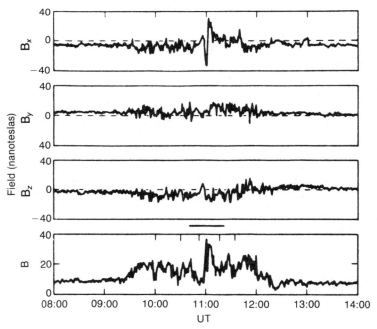

Figure 6.3 Overview of magnetic field observations near comet Giacobini-Zinner. The measurements show large fluctuations in total magnetic field (B) and in the components. At closest approach (11:02 UT), the total field shows a maximum value and the component in the direction away from the sun (B_x) reverses sign as Alfvén's model predicts. (Courtesy of E. J. Smith, NASA-Jet Propulsion Laboratory)

the magnetic field was observed to reverse polarity near this time. The observed reversal occurred when the spacecraft passed from one magnetic lobe through the 1000-kilometer-thick current sheet and into the opposite magnetic lobe. The observations showed the plasma tail to be about 10,000 kilometers across and to contain field strengths (magnetic flux densities) as high as 60 nanoteslas (nT). To put this number in perspective, the earth's magnetic field at the equator has a flux density of about 30,000 nanoteslas.

The plasma behavior followed a specific pattern. Far from the comet, the plasma showed typical solar-wind values. It flowed at a speed of about 500 kilometers per second, with an electron temperature of 250,000 K and a density of 5 ions per cubic centimeter (see Figure D.1 in Appendix D). Although there were some variations, the trend observed as the spacecraft moved toward its closest approach was toward higher densities, lower temperatures, and lower flow speeds. The pattern reversed after the closest approach, and these quantities returned to the solar-wind values. Near the closest approach the electron densities exceeded 600 ions per cubic centimeter, the temperature was about 15,000 K, and the flow speed was less than 30 kilometers per second.

The observed cometary ions were mostly water-group ions, H_2O^+ and H_3O^+; in addition, CO^+ and HCO^+ were probably detected. The predominance of water-group ions supports Whipple's dirty-snowball model of cometary nuclei.

High-energy ions were observed over a large region extending from 1 day before closest approach to $2\frac{1}{2}$ days afterward (see Figure D.2). A reduction in their number was seen near the time of closest approach, when the spacecraft was in the cold, dense plasma tail. Most of the energetic ions resulted when water-group molecules were ionized, picked up by the solar wind's magnetic field (and hence called "pickup ions"), and accelerated by the solar-wind flow. Other acceleration processes are probably required to explain the observations at the highest energies.

Plasma waves were detected almost continuously within 2 million kilometers of the nucleus of comet Giacobini-Zinner (see Figure D.3). The wave levels increased as the spacecraft approached the nucleus. The waves are probably generated by the strong interaction of the newly created molecular ions (such as H_2O^+) with the solar wind via its magnetic field. The same interaction produces the energetic ions described just above. The same experiment that measured the plasma waves could also measure the impact of dust particles on the spacecraft. The spatial extent of these collisions corresponded to the extent of the visible coma. The rate was as low as anticipated.

A *bow wave* (perhaps a shock) or transition region was detected by several experiments at approximately 130,000 kilometers from the nucleus on both the inbound and outbound legs. The wave or transition region is thick (roughly 40,000 kilometers) and functions to decelerate the solar wind from supersonic to subsonic flow. Bow waves are common in the solar system because the supersonic-to-subsonic transition is required to allow the solar wind to flow smoothly around an obstacle (a comet, the earth, whatever). Because the interaction of the solar wind with the comet (unlike other obstacles in the solar system) takes place over a huge volume of space, this

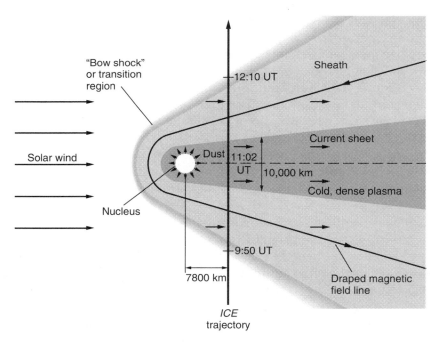

Figure 6.4 Summary of results of the *ICE* encounter with comet Giacobini-Zinner. UT times along the trajectory refer to the encounter date, September 11, 1985.

bow wave is actively being studied; many details are yet to be settled.

A summary of the results obtained from in situ measurements of comet Giacobini-Zinner is shown in Figure 6.4. Generally, plasma waves and energetic ions were found over the entire area studied. Many features of Alfvén's model of plasma tail formation were confirmed, and the chemical composition results supported Whipple's model of the nucleus. The surprises included the large distances over which plasma wave activity and energetic ions were detected and the large short-term variations (possibly turbulence?) observed in most physical parameters.

The *ICE* measurements provided a first glimpse of in situ cometary measurements some six months before the Halley Armada began its work. Shortly after the *ICE* encounter, comets Giacobini-Zinner and Halley were within 2 degrees of each other in the sky. This historic changing of the guard is shown in Figure 6.5.

The 1982 Recovery of Comet Halley

The recovery of comet Halley in 1982 could not have been duplicated by amateur astronomers. It was accomplished with the 200-inch Hale telescope (on Mount Palomar) outfitted with a sophisticated electronic camera, which is a prototype of a camera being flown on the Hubble Space Telescope. The astronomers G. Edward Danielson and David Jewett (Figure 6.6) were using this equipment to make images of the sky in the direction of the constellation Canis Minor, hoping that comet Halley would show up on the image. On the night of October 16, 1982, they were rewarded. The comet appeared as a faint starlike image (Figure 6.7) with a magnitude of roughly 24.5. How did Danielson and Jewett know that the tiny, faint dot was the comet and not a star or, worse, a blemish? They did not know at first. But they continued to observe the same region of the sky over the course of several nights, and each time they observed the tiny dot, they saw it moving slightly with respect to the background stars. The image was right where comet Halley should have been and was moving just as Halley

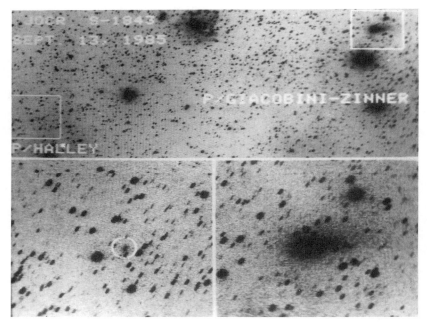

Figure 6.5 The changing of the guard as comets Giacobini-Zinner and Halley appear on the same wide-field image taken on September 13, 1985, just two days after the historic encounter. (JOCR photograph by E. Moore)

Figure 6.6 The recoverers of Halley's comet, David C. Jewett (left) and G. Edward Danielson, examine the image obtained on October 16, 1982. (California Institute of Technology photograph)

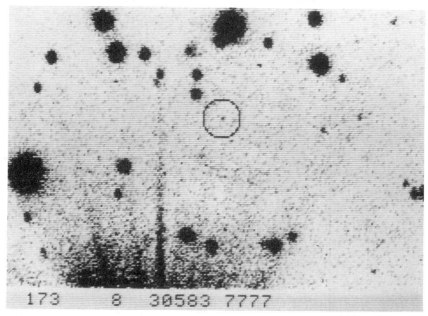

173   8   30583 7777

Figure 6.7 The recovery image of Halley's comet (circled) obtained on October 16, 1982, by means of an advanced electronic detector system and the 200-inch Hale telescope at Palomar Observatory. (California Institute of Technology photograph)

should have been moving, according to the calculations of the celestial mechanicians. Those two facts were very strong evidence that the dot was comet Halley.

We can see how observing technology has progressed since Halley was first discovered when we note that on the 1759 pass, comet Halley was discovered about a month before it passed perihelion; in 1835 it was rediscovered three months before perihelion; in 1910 it was rediscovered nine months before perihelion; and in 1982 it was rediscovered more than three years before perihelion. When it was rediscovered in 1982, comet Halley was beyond the orbit of Saturn, more than 10 astronomical units from the sun, a record for the observation of any comet (Figure 6.8).

Planning for Space Missions to Study Comet Halley

As the 1985–1986 apparition of Halley's comet drew near, enthusiastic planning in the space science community acceler-

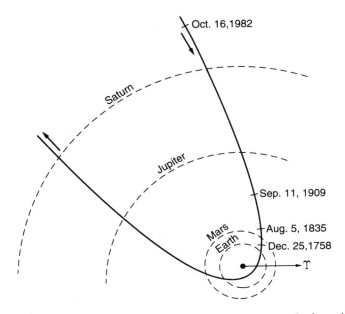

Figure 6.8 The recovery distance for Halley's comet marked on its orbit projected into the earth's orbital plane. The improvement over time is clearly seen. The ♈ marks the direction to the vernal equinox.

ated. Students of comets argued that spacecraft should be sent to the comet so that the nucleus could be photographed, the chemical composition could be measured in situ, and other measurements could be made that would help resolve many outstanding issues in our understanding of comets. An extensive program of supporting ground-based observations was contemplated.

The plans for Halley took an ironic turn about five years before the space missions were to occur. Ambitious *flyby missions* were being planned by NASA, the European Space Agency (ESA), Japan, and the Soviet Union. Only the United States failed to approve a mission dedicated to Halley's comet. This failure of the NASA planning process was partially compensated for by the beautifully improvised *ICE* mission to comet Giacobini-Zinner and by NASA's leadership in the International Halley Watch. The intensive study of comets in 1985–1986 became the study of two comets, Giacobini-Zinner and Halley.

The United States's involvement with comet Halley received a disastrous blow in early 1986. One of the experiments lost in the *Challenger* disaster was a spectrograph designed to observe Halley's comet while it was near the sun. And because further shuttle flights were then cancelled, extensive observations of the comet planned for early March 1986 by the *Astro 1* mission—a trio of instruments designed to study the ultraviolet light from celestial objects augmented by a wide-field camera—were also lost.

The Halley Armada

Five spacecraft besides *ICE* were sent to intercept Halley's comet. Some details and an indication of imaging capability are shown in Table 6.1. A schematic presentation of some of the same information is given in Plate 9.

The problem in celestial mechanics that had to be solved if Halley's comet was to be intercepted was relatively straightforward. Researchers recognized that a spacecraft could deliver its maximum payload at the comet by intercepting it as it crossed the earth's orbital plane at a location called the node of the orbit. The comet at the time in question (March 10, 1986) was moving from above the earth's orbit (north) to below the earth's orbit (south), at what is called the descending node. Thus the spacecraft encounters are clustered around this date, as we see in Table 6.1. Figure 6.9 shows the orbital geometry of *Giotto*'s trajectory and the descending node concept. The community of cometary scientists enjoyed exceptionally good fortune in that all these comet missions were launched successfully and nearly all of the roughly 50 experiments on the spacecraft functioned properly.

The two Soviet *Vega* spacecraft were originally intended to go to the planet Venus. The missions followed an orbit that took them first past Venus, where they dropped off instrument packages before they proceeded to the comet. The dates of the Venus flybys were June 11, 1985, for *Vega 1* and June 15, 1985, for *Vega 2*. "*Vega*" incorporates the first two letters of the Russian name for Venus (*Venera*) and the Cyrillic spelling of Halley (*Galley*).

Table 6.1 Space Missions to Comets Giacobini-Zinner and Halley

| Spacecraft | Sponsor | Comet | Closest approach | | | Date of closest approach | Flyby speed (km/sec) | Imaging system for nucleus |
			Approximate distance (km)	Orientation	Launch date			
ICE	NASA	Giacobini-Zinner	7,800	Tailward	Dec. 22, 1983[a]	Sep. 11, 1985	21	No
Vega 1	USSR	Halley	8,890	Sunward	Dec. 15, 1984	Mar. 6, 1986	79	Yes
Suisei	Japan	Halley	151,000	Sunward	Aug. 18, 1985	Mar. 8, 1986	73	No
Vega 2	USSR	Halley	8,030	Sunward	Dec. 21, 1984	Mar. 9, 1986	77	Yes
Sakigake	Japan	Halley	7,000,000	Sunward	Jan. 8, 1985	Mar. 11, 1986	75	No
Giotto	ESA	Halley	596	Sunward	July 2, 1985	Mar. 14, 1986	68	Yes
ICE	NASA	Halley	28,000,000	Sunward	Dec. 22, 1983[a]	Mar. 25, 1986	65	No

[a] Final lunar encounter and effective launch to the comet; launch from earth was Aug. 12, 1978.

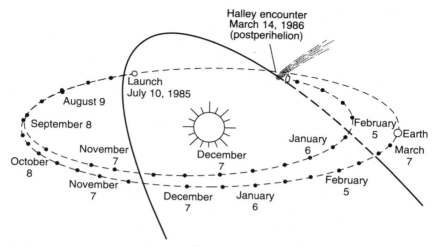

Figure 6.9 *Giotto*'s trajectory from launch on July 10, 1985, to its encounter with Halley's comet on March 14, 1986. (Courtesy of R. Reinhard, European Space Agency)

Each *Vega* spacecraft carried an extensive instrument complement. A primary goal of the mission was to image the nucleus, and the spacecraft carried camera systems for this purpose. In addition, the spacecraft were outfitted with spectrometers to record emissions from the nucleus and coma, including thermal emission from the nucleus. The spacecraft also carried several experiments designed to measure cometary dust—its chemical composition, its amount, and the distribution of its particles of various sizes.

Another group of instruments measured the composition of neutral and ionized gases and the energy spectrum of electrons; that is, the numbers of electrons with energy in specified intervals. Finally, instruments measured the magnetic field and plasma waves around the comet. Plasma physics is often thought to be somewhat esoteric, but its importance for an understanding of the cosmos cannot be underestimated. To place the matter in perspective, roughly half of the experiments sent on spacecraft to comets in 1985 and 1986 dealt with plasma physics.

Japan's contribution to Halley studies consisted of two spacecraft. *Sakigake* translates as "pioneer" or "forerunner," with the connotation of risk-taking. This spacecraft was Japan's first mission into interplanetary space and was largely an engineering test. *Suisei,* the name of the second spacecraft, means "comet." The two spacecraft are nearly identical but they carried different scientific experiments. The instruments on *Sakigake* were designed to measure basic properties of the solar wind near the comet, radio emission from the coma, and plasma waves. *Suisei* carried an ultraviolet imaging system to observe the hydrogen cloud around the comet and ions in the solar wind and from the comet.

The *Giotto* spacecraft, launched by the European Space Agency, was named after the Italian Renaissance painter Giotto di Bondone, who depicted the star of Bethlehem as a comet in one of the Arena Chapel frescoes in Padua (Plate 4), as we saw in Chapter 1. *Giotto* was ESA's first interplanetary mission and its first scientific spacecraft launched by an *Ariane rocket,* an unmanned liquid-fueled rocket capable of placing scientific payloads in earth orbit.

Giotto carried an extensive array of scientific instruments. A paramount goal was the imaging of the nucleus and inner coma from a distance of a few hundred kilometers. The resolution of the imaging system would be 11 meters at a distance of 500 kilometers. *Giotto's* instruments measured the composition of neutral atoms, ions, and dust particles. Additional experiments measured the amount of dust and the distribution of its particles. Photometric measurements determined the brightness of the coma. A suite of plasma instruments measured the properties of ions and electrons in a variety of energy ranges, and a magnetometer measured the comet's magnetic field. The broad goal of these plasma experiments and those on the other spacecraft was to study the interaction of the solar wind with comets.

The problem of damage to the various spacecraft and their scientific instruments by the dust near the nucleus of comet Halley was significant and was solved in a variety of ways. *Suisei* and *Sakigake* solved the problem by avoiding the dust. Models predicted a very low hazard at distances of roughly 200,000 kilometers. Some parts of the *Vega* spacecraft required protection, but the existence of two spacecraft provided a measure of redundancy. The close-encounter distance of *Giotto* demanded special precautions.

The flyby speeds were high because all the spacecraft's orbits took them around the sun in the same direction as the planets, whereas comet Halley moves in the opposite (retrograde) direction. The encounter was rather like freeway travel in the wrong lane. Encounter speeds were roughly the same for all the spacecraft: *Giotto*'s, for example, was 68 kilometers per second. At this speed, a dust particle with a mass of 0.1 gram could penetrate an aluminum sheet 8 centimeters thick.

Giotto's solution to the dust penetration problem was a double bumper shield. Figure 6.10 shows the shield and the rest of the spacecraft. The bumper consists of a thin front sheet, a large gap, and a thick rear sheet. A dust particle striking the front sheet would vaporize and expand into the gap. When it reached the rear sheet, its energy would be dissipated over a large area. Estimates indicated that the shield could withstand an impact from a dust particle with a mass of roughly 1 gram. An additional problem was that the impact of a much smaller particle on the rim of the shield could cause the spacecraft to wobble and lose contact with earth. Such an event occurred 14 seconds before closest approach, and data were received only intermittently for the next 32 minutes.

At the high flyby speeds, dust impacts were clearly hazardous to the spacecraft. On *Giotto,* the capabilities of several experiments were degraded or the instruments were knocked out completely. The dust impacts caused some instruments to fail on the *Vega*s, and the power available from the solar cells was reduced by about 50 percent. Even *Suisei* was hit near closest approach by two dust particles with masses of several milligrams, despite its distance of 151,000 kilometers from the nucleus. Only *Sakigake* and *ICE* appear to have made their voyages unscathed.

Giotto and the *Vega*s are still orbiting through the solar system and are functioning spacecraft. Any of them could be retargeted to study another comet or an asteroid. The *Giotto* Extended Mission (GEM) has been approved by ESA, and reactivation of the spacecraft is scheduled. The target will be comet P/Grigg-Skjellerup in July 1992.

The Pathfinder project, a part of the comet Halley encounters, was a beautiful example of international cooperation. Getting *Giotto* to pass within about 500 kilometers of Halley's nucleus required exceptional targeting accuracy. Traditional

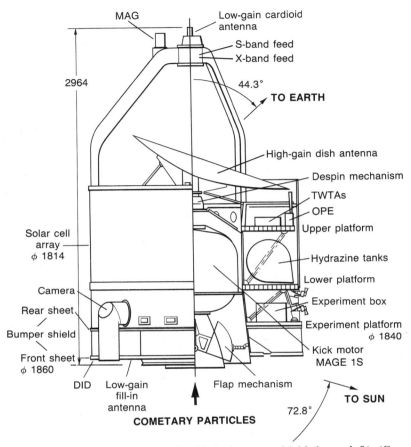

Figure 6.10 The *Giotto* spacecraft with its bumper shield (lower left). (Courtesy of R. Reinhard, European Space Agency)

ground-based measuring techniques were so lacking in accuracy that their use might have caused the spacecraft to pass on the dark side of the nucleus or even to collide with it. NASA's part of the project was to determine the positions of the *Vega* spacecraft by carrying out very long baseline interferometry (VLBI) with its Deep Space Network; the *Vega*s were to determine the position of the comet at their times of closest approach on March 6 and 9, 1986. The positional data had to be processed and communicated rapidly to ESA for *Giotto*'s encounter on March 14, 1986. The successful result permitted a final flyby distance of 605 kilometers to be targeted, with an uncertainty of 40 kilometers. The flyby distance actually achieved was 596 kilometers. The remarkable success of this international project shows what can be done when nations cooperate.

The direct explorations carried out by the various spacecraft are quite impressive, but our progress toward an understanding of the physics of comets does not rest on these missions alone. Important data have been received from other sources. Spacecraft in earth orbit, such as the *International Ultraviolet Explorer (IUE)*, the *Solar Maximum Mission (SMM)*, and the *Dynamics Explorer 1*, have also contributed. The *Pioneer Venus Orbiter* in Venus's orbit has contributed important data, as have numerous rocket flights and the *Kuiper Airborne Observatory*. Last, but emphatically not least, are the ground-based networks of the International Halley Watch. Our understanding of comets will come from a synthesis of the data from all sources in combination with theoretical insight.

Focus on Comet Halley

Wide-Scale Phenomena

Plasma tail phenomena in comet Halley began in mid-November 1985, when a short tail appeared, disappeared, and reappeared a few times (Figure 6.11). By early December 1985 the plasma tail was a permanent fixture.

One of the goals of extensive imaging was to record the sequence of events expected when the entire plasma tail detaches from the comet, floats away, and is replaced by a new tail. A spectacular disconnection event took place on January 9, 10, and 11, 1986; and many other such events were recorded during the 1985–1986 appearance of Halley's comet.

Comet Halley passed perihelion on February 9, 1986. One of the first images obtained after perihelion is shown in Figure 6.12. The photograph, taken on February 22, 1986, clearly shows multiple dust tail structures, an antitail—that is, a tail-like structure pointing toward the sun—and a disconnected plasma tail.

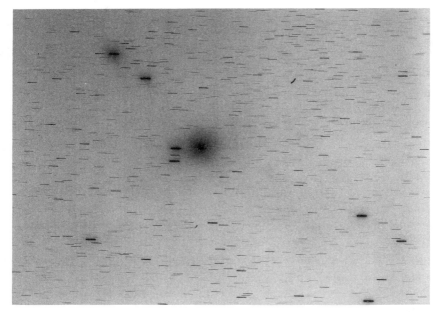

(a)

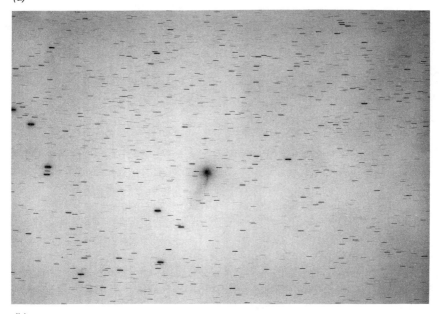

(b)

Figure 6.11 Comet Halley in November 1985. *(a)* On November 12 the comet is essentially a featureless, spherical coma with no tail. (Photograph by H. Meusinger and K. Mau, Karl Schwarzchild Observatory, Tautenburg, Germany) *(b)* On November 13 a plasma structure is clearly seen. (Photograph by L. Kohoutek, Calar Alto Station, Spain, of the Max-Planck-Institut für Astronomie, Heidelberg)

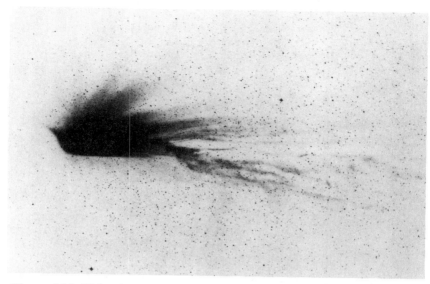

Figure 6.12 This photo of comet Halley taken February 22, 1986, shows structures in the dust tail, an antitail, and a disconnected plasma tail. (U.K. Schmidt Telescope in Australia, © 1986, Royal Observatory, Edinburgh)

The period from March 6 to 14, 1986, was the time of the spacecraft encounters and an exciting time for the ground-based observers. The combination of ground-based observations and in situ measurements can help researchers unravel some of the complex issues of cometary studies. The large-scale structure of comet Halley, including a disconnection event on March 9 and 10, 1986, is shown in Figure 6.13.

The best period for naked-eye viewing was in late March and early April 1986. An image taken during this period is seen in Figure 6.14. (See also Plates 1, 2, 7, 8, and 11.) Major tail structures continued to be observed through April 1986. Then in early May large-scale plasma tail phenomena ceased, but dust features, probably the "neckline" structures (Figure 6.15), persisted into July.

Thus large-scale plasma phenomena in comet Halley lasted approximately six months and were covered extensively from the ground. A major occurrence was the direct measurement of the kind of plasma phenomena observed at comet Giacobini-Zinner, by the armada of spacecraft sent to Halley.

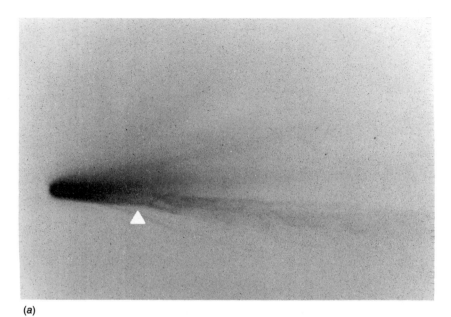

(a)

(b)

Figure 6.13 A spectacular disconnection event in Halley's comet during the time of the spacecraft encounters: *(a)* March 9, 1986, and *(b)* March 10, 1986. (U.K. Schmidt Telescope in Australia, © 1986, Royal Observatory, Edinburgh)

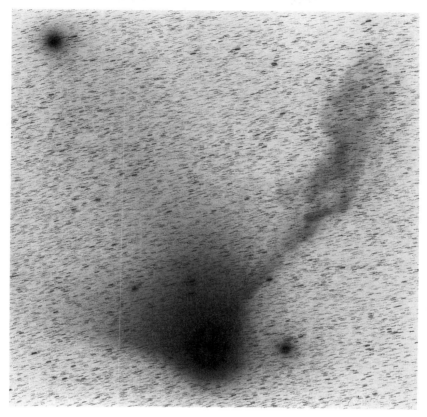

Figure 6.14 A dust tail (left) and a complex plasma tail (right) are seen in comet Halley on April 12, 1986. (Courtesy of F. D. Miller, University of Michigan, CTIO)

Figure 6.16 summarizes the plasma results from comet Halley. As a crude approximation, the geometric scales for plasma structures in comet Halley are those observed for comet Giacobini-Zinner but scaled up by a factor of 7. The bow shock was found about 1 million kilometers from the nucleus and had a thickness of about 100,000 kilometers. The plasma-flow speed was measured by several spacecraft and varied from the solar-wind speed far from the comet to 10 kilometers per second near the nucleus. Abundances of various ions were measured, and H_3O^+ was found to be the most abundant ion near the nucleus. Extensive energetic ions were observed in an extended region surrounding the nucleus.

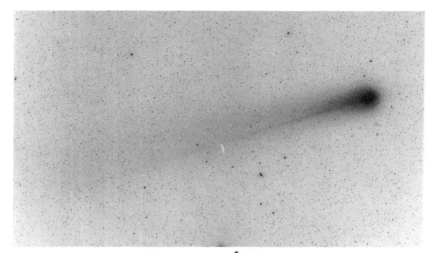

Figure 6.15 Comet Halley on May 7, 1986. The thin, stable dust feature is probably the neckline structure. (Courtesy of F. D. Miller, University of Michigan, CTIO)

The interaction of the comet and the solar wind is beautifully illustrated by data from the ion analyzer on board *Giotto* (see Plate 11). The flowing plasma is composed of two distinct mass groups: (1) the pickup ions in the mass group 12–22 atomic mass units (AMU), which show up at an ion energy of about 3×10^4 electron volts (eV)—note that water ions, H_2O^+, would be 18 AMU (because H is 1 AMU and O is 16 AMU); and (2) the solar-wind protons (1 AMU for H^+), which show up at 10^3 eV. Some transitions are marked. Outside of point 1 the velocity of the solar-wind protons is at the expected solar-wind speed and the pickup ions are at the energy expected for water-group ions. The velocity of all ions shows a steady decrease, as decreasing ion energy in Plate 10 indicates. At point 1 the flow speed decreases, the flow is deflected away from the sun-comet line, and the temperature or random motion of the solar-wind and pickup ions increases, as the increased width (in energy) of the distributions indicates. This is the diffuse bow shock detected at 1 million kilometers from the nucleus with a width of about 100,000 kilometers.

The second transition (point 2) occurs at about 500,000 kilometers from the nucleus. Here the flow is further deflected

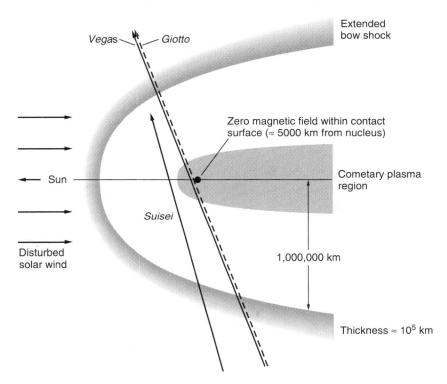

Figure 6.16 Summary of plasma results at Halley's comet, drawn approximately to scale. The lines representing the paths of *Giotto* and the *Vegas* are displaced for clarity.

from the sun-comet line and the width of the distribution of the solar wind's proton energy increases further, but the pickup ions show a decrease in energy width and a population with lower energy appears. The speed of the pickup ions decreases rapidly as the nucleus is approached, and at a distance of approximately 80,000 kilometers the flowing plasma is depleted.

The basic physics of ion pickup is as follows. A newly created cometary ion is essentially at zero velocity with respect to the typical solar-wind flow of 400 kilometers per second. The ion, however, cannot cross the magnetic field lines in the solar wind. Therefore, the solar wind picks up the ion, and the wind-flow speed, v, becomes the speed at which the ion spirals around the field line while also being carried along at the solar-wind speed. Thus, when the ion is spiraling in the same direction as the flow, the speed can be twice the solar-wind speed. The ion's energy is proportional to its speed squared, and the energy can be larger by a factor of 4 than it would be if it were moving at the solar-wind speed. When the motion is not perpendicular to the field lines, the maximum energy is reduced by a factor of $(\sin \psi)^2$ where ψ is the angle between the direction of motion and the field lines.

The magnetic field observations showed the bow shock wave, the draped-field-line configuration, a magnetic pile-up region, and a central cavity free of magnetic field (see Figure D.4). The magnetically observed bow shock is at essentially the distance observed by the other experiments on several spacecraft. The magnetic pileup region, a region with magnetic field strength enhanced over the solar-wind value, was measured at distances within 135,000 kilometers from the nucleus when the spacecraft was inbound and at distances within 263,000 kilometers from the nucleus when the spacecraft was outbound; the fields reached values of 57 and 65 nanoteslas in the respective regions. The field-free cavity (see Figure D.5) has a diameter or width of 8500 kilometers. The cavity is formed by the outflowing pure cometary plasma, and the boundary is the *contact surface* that separates the pure cometary plasma near the nucleus from the mixed cometary and solar-wind plasma farther out.

The magnetic field measurements made by the spacecraft during Armada Week are important with respect to the disconnection events that were observed in wide-field imaging (Figure 6.13, for example); see Chapter 4.

Finally, cometary scientists are interested in the dimensions of the interaction region in which Halley's comet influences the solar wind. The measurements made by *Sakigake* clearly showed the influence of the comet some 7 million kilometers sunward of the nucleus, and measurements made by *ICE* some 28 million kilometers (or 0.2 astronomical unit) sunward of the nucleus may have shown the comet's influence. Cometary effects observed at the greater distance are compatible with the measurements made at comet Giacobini-Zinner, scaled up by a factor of 7. Despite some uncertainty about the largest dimension, a Halley-size comet has a large interaction region, roughly 0.1 astronomical unit or greater.

The Hydrogen Cloud and the Coma

Because the hydrogen cloud is so immense, it is best observed from afar. An instrument on board the *Pioneer Venus Orbiter* had an excellent view of Halley's comet in early 1986. The hydrogen cloud as observed in the light of the hydrogen Lyman-alpha emission is shown in Plate 5; for comparison, the white circle is the size of the sun's disk at the same distance. Analysis of the data on the hydrogen cloud has established the rate of total water production in early 1986, including a peak value of 1.6×10^{30} molecules per second (or about 50 tons per second) in mid-February.

Observations and measurements of the coma, the roughly spherical cloud of dust and gas surrounding the nucleus, were extensive. Photographs that reveal structures within the coma are shown in Figure 6.17, and sample spectra are shown in Figure D.6. We begin with the gas. About 80 percent of the inner coma was found to be water or water-group molecules (H_2O, OH, H_3O^+, H_2O^+, OH^+). About 10 percent was determined by a terrestrial rocket experiment to be carbon monoxide (CO). Carbon dioxide (CO_2) accounted for about 3.5 percent and polymerized formaldehyde [$(H_2CO)_n$] for probably a few percent. The remainder consisted of trace substances, the most abundant of which we expect to be methane (CH_4) and ammonia (NH_3). These values apply only to Halley, but they probably are typical of a water-dominated comet. They probably do not apply to a fresh (new) comet and perhaps not to the bulk of comets.

The discovery of polymerized formaldehyde is based on mass peaks (see Figure D.7) in the heavy-ion mass spectrum at atomic mass unit values of 45, 61, 75, 91, 105, and probably 121. These can be considered segments of polymer chains of formaldehyde [$(H_2CO)_n$]. The different end caps to the chain account for the differing separations and the widths of the peaks. While the identification of polymerized formaldehyde is entirely plausible, the possibility that other compounds could mimic the mass peaks assigned to this molecule should be kept in mind. Thus it is best to think of polymerized formaldehyde as a likely constituent and as a surrogate for other possible complex organic compounds.

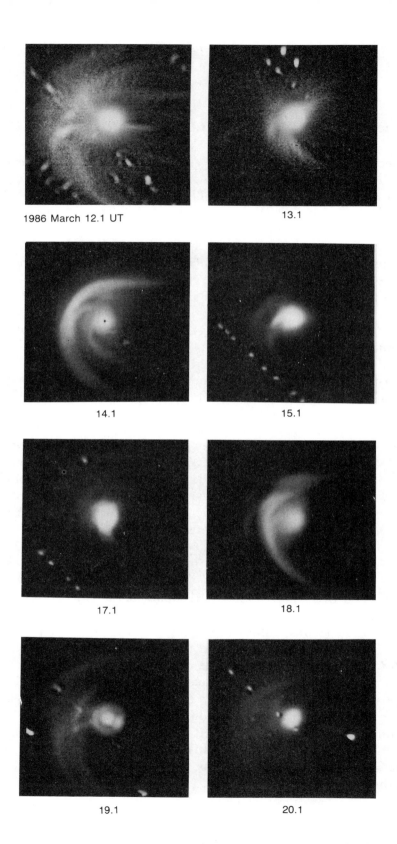

1986 March 12.1 UT

13.1

14.1

15.1

17.1

18.1

19.1

20.1

The unequivocal discovery of organic molecules in comets is a clue to their origin and is of considerable interest to students of comets. The possible discovery of polymerized formaldehyde generated some excitement, too. Another measurement of interest to researchers who view comets as a possible source of the water in the earth's oceans is the ratio of deuterium to hydrogen (D/H) in the water vapor. The value is 2×10^{-4}, and the uncertainty is fairly large.

The density of water vapor as a function of distance from the nucleus is shown in Figure D.8. At 1000 kilometers from the nucleus, the density is about 5×10^7 molecules per cubic centimeter and the expansion speed is 0.9 kilometer per second. These values tell us that 7×10^{29} molecules per second cross a sphere 1000 kilometers in diameter centered on the nucleus, suggesting a total gas production rate of about 7×10^{29} molecules per second.

Dust detectors on the *Vega*s and *Giotto* made measurements of number densities (number of particles per cubic centimeter) and the mass distribution of dust particles. The number of small grains (mass $\sim 10^{-17}$ gram) was larger than ground-based observations had led scientists to expect. The fraction of smaller particles increased with distance from the nucleus, a fact interpreted as indicating the existence of some large, fragile dust particles that fragment on their journey outward. They could be quite similar to Brownlee particles—small porous particles a few microns in size that have been collected in the upper atmosphere and are believed to be cometary in origin (see Figure 8.2). The total density at (say) 8100 kilometers is 3×10^{-6} particle per cubic centimeter, for particles with $m \geq 1.5 \times 10^{-13}$ gram; the mass distribution is shown in Figure D.9. The expansion speeds are 0.5 kilometer per second for a 10^{-8}-gram particle and 0.75 kilometer per second for a 10^{-12}-gram particle. The lighter particles are expected to be carried along at higher speeds by the gas flow.

The dust composition measurements made by *Giotto* and the *Vega*s disclosed three kinds of dust particles. The first kind

Figure 6.17 These images of comet Halley in March 1986 have been computer-enhanced to show the structure of the coma. (Courtesy of S. Larson, International Halley Watch, University of Arizona)

shows a silicate composition roughly similar to a type of meteor known as a carbonaceous chondrite. The second shows a composition consisting of only the light elements carbon, hydrogen, oxygen, and nitrogen—dubbed the CHON particles. The third type is a mixture and can be described as silicate particles enriched with the lighter elements; these make up the majority of the particles. Examples of the mass spectra are shown in Figure D.10.

The third type, composite particles, resemble the particles produced by J. Mayo Greenberg in his laboratory at Leiden. Basically they are composed of a refractory core—that is, a core with a high melting point—with an outer coating of complex organic materials. The ratio of carbon to silicon is higher than in any known class of meteorite. Nitrogen is present in the particles with high carbon/silicon ratios, a fact entirely consistent with the presence of carbon and nitrogen atoms as part of organic molecules.

A ground-based observation of jets in the coma is shown in Plate 12. The jets were imaged in the light of the cyanogen molecule and image-processed to show detail. These jets do not coincide with the dust visible in photographs of the nucleus. The relation of the cyanogen jets to the CHON particles is also unclear.

Variations in Brightness

The record of variations in the brightness of Halley's comet at its last apparition does not refer solely to the nucleus or the coma. Figure D.11 shows the brightness variation over the entire apparition (with some interpretation), and Figure 6.18 shows the behavior around perihelion. (We will discuss brightness in connection with the nucleus and nuclear activity separately.)

In general, the following semiqualitative observations can be made: (1) The light curve is highly asymmetrical and is much brighter after perihelion. (2) Significant short-term (about 1–2 days) variations are present; mechanisms capable of producing large changes or outbursts may be required to explain these variations. (3) The onset of coma activity appears to have occurred at roughly 6 astronomical units. (4) The brightness var-

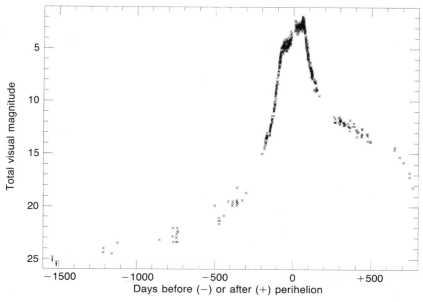

Figure 6.18 Brightness variation of Halley's comet around perihelion. This light curve is an updated version of the one published in *Astronomy and Astrophysics* 187 (1987): 560. Perihelion occurred on February 9, 1986. Data from *International Comet Quarterly* archive. (Courtesy of D. W. E. Green, Harvard-Smithsonian Astrophysical Observatory)

iations cannot be simply represented by any of the traditional formulas.

The Nucleus

Observations of the nucleus began with the recovery of the comet in October 1982; the recovery observation is shown in Figure 6.7. The light curve from this first observation to early 1991 is shown in Figure D.11. From heliocentric distances of 11 to approximately 6 astronomical units, the brightness variations *roughly* follow the law expected for a bare nucleus without coma activity. The onset of measurable coma activity began when the comet was about 6 astronomical units from the sun. Having said this, we must point out that the brightness variations at great distances (Figure D.11) may require explanations other than a rotating nucleus. If some of the variations in measured brightness are not attributable to rotation or uncertainty of measurement, other mechanisms may have to be identified to explain them. Rough estimates of the type of variations one might expect are as follows. Direct observations of the nucleus indicate that it is an irregular object, and its observed surface area could vary by as much as a factor of 2. If the reflectivity of the nucleus were constant over the entire surface, then the brightness could vary by the same factor of 2 because the nucleus rotated in relation to the observer. It is difficult to measure brightness at the very low levels of Halley's comet in 1982 to 1984, and uncertainties of several tenths of magnitudes are expected. Brightness variations larger than those expected from the rotation of the nucleus must be attributed to other causes.

One other possible cause of brightness variations is an "outburst" mechanism involving transitions between different phases of ice (see Chapter 3), but as we will see, there is some doubt about its applicability. If ice is formed by condensation on a surface at very low temperatures, the ice is in the amorphous form; at these low temperatures, the water molecules do not have sufficient energy to change the ice into a configuration that minimizes energy. As the temperature increases, the energy of the molecules increases and regular configurations can be achieved. The ice assumes first a cubic crystalline form and then the common hexagonal crystalline form. Both of these transitions are exothermic—that is, energy is released when they occur. A sudden release of energy could cause a blow-off of surface material and thus an increase in brightness. These phase transitions occur at temperatures of about 136 K and 173 K, respectively. The applicability of this mechanism to Halley's comet is questionable because these mechanisms do not occur in reverse; in other words, when the temperature of hexagonal ice is lowered from 173 K to 136 K, it does not revert to cubic ice, but remains hexagonal ice. In addition, Halley's comet has made many passes through the inner solar system, and its subsurface layers may have been heated above the transition temperatures. This mechanism could apply to new comets, however.

The second "outburst" mechanism could result from electrical charging of the night side of the nucleus—that is, the side away from the sun. On the side toward the sun, the day side, electrons and protons from the solar wind strike the surface in equal numbers and no electrical charge results. The situation is different on the night side (Figure 6.19). The random speeds of electrons in the solar wind are much higher than those of protons, so more electrons than protons strike the dark surface. A charge builds up to an equilibrium value—that is, the charge itself attracts protons to the dark side and an equilibrium is established. The charge can cause dust on the surface to levitate. A sudden change in solar-wind conditions could cause dust to blow off and produce an increase in brightness.

By the time measurable coma production begins—that is, when the comet is about 6 astronomical units from the sun—these exotic charging mechanisms can no longer operate, since solar-wind particles can no longer reach the nucleus. Then the

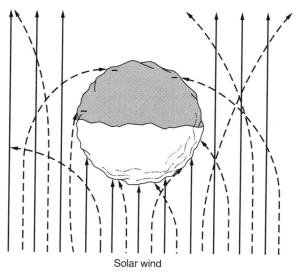

Solar wind

Figure 6.19 Mechanism to charge the night side of a cometary nucleus far from the sun. The solid lines represent the paths of protons, the dashed lines the paths of electrons. Because the random speeds of electrons in the solar wind are high, the paths of electrons are curved and some deposit a charge (−) on the night side of the comet. This process continues until the negative charge on the comet attracts sufficient protons to stop the buildup.

brightness or coma evolution more or less follows the standard
model of ice sublimation. This is shown by the variation in
magnitude when the comet is less than 6 astronomical units
from the sun (Figure D.11).

Our specific knowledge of the nucleus comes from observations by instruments on board *Giotto* and the *Vega*s. There

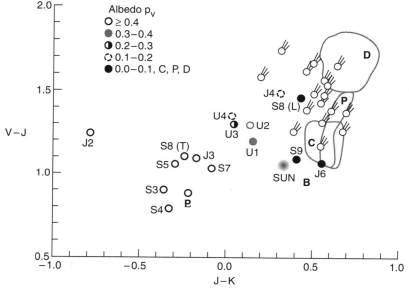

Figure 6.20 Infrared (VJK) color-color diagram for asteroids, comets, satellites, and other objects in the solar system. The quantitites V, J,
and K are the mangitudes of the various objects measured
through filters that pass radiation with wavelengths in the infrared between 12,500 and 22,000 angstroms. The differences
V-J and J-K, called colors, tell us the relative reflectivities of the
objects at these infrared wavelengths. The satellites of the planets
are labeled with the first letter of the planet's name and the
number of the satellite; for example, U3 is the third satellite of
Uranus. The L or T after the satellite designation indicates the
leading or trailing side of the satellite, relative to its orbital motion, and ℗ is the planet Pluto. The comets are shown with the
comet symbol (⚹). The asteroids are shown by type, with fields
marked by B, C, D, and P. The comets all fall into the low
albedo region of the diagram, mostly less than 0.1, and the colors are consistent with dirty ice colored by dark soils. (Courtesy
of W. K. Hartmann, Planetary Science Institute)

were some surprises. Scientists expected a roughly spherical nucleus, a radius of about 2.5 kilometers, and albedos (fraction of light reflected) of up to 0.6. The actual nucleus was larger, darker, and substantially more irregularly shaped than had been expected. In retrospect, the dark nucleus should have been expected on the basis of infrared observations of comets, satellites, and asteroids (see Figure 6.20).

Images of the nucleus are shown in Figures 6.21 through 6.24. The irregular shape is obvious. Descriptions of the shape are somewhat subjective; the *Vega* investigators describe it as a peanut and the *Giotto* investigators as a potato. Figure 6.25 shows two representations with dimensions; the nucleus is approximately 16 kilometers long and 8 kilometers across at the widest point. The surface is locally irregular with a variety of

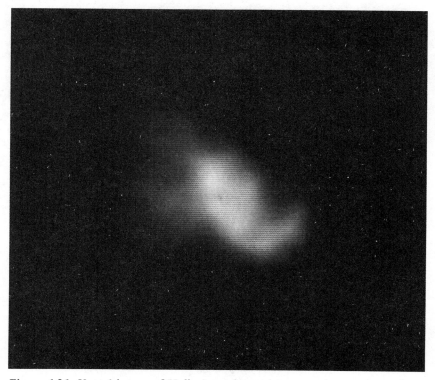

Figure 6.21 *Vega 1* image of Halley's nucleus taken near closest approach at a range of 8904 kilometers. (Courtesy of E. Merenyi, Lunar and Planetary Laboratory, University of Arizona)

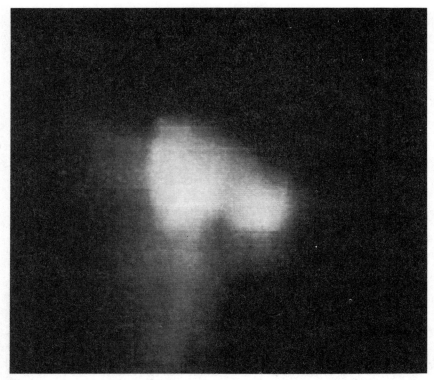

Figure 6.22 *Vega 2* image of Halley's nucleus taken near closest approach at a range of 8031 kilometers. (Courtesy of E. Merenyi, Lunar and Planetary Laboratory, University of Arizona)

features that can be called craters, hills, slopes, and the like. Some of them are labeled in Figure 6.24*b*.

With a fraction of light reflected (or albedo) of about 3 percent, the nucleus is one of the darkest objects in the solar system. The nucleus is darker than coal or black velvet. The low albedo is probably due to the presence of a dust crust. This view is consistent with a surface temperature of 330 K determined by infrared measurements. This temperature is close to that of a slowly rotating black body at 0.8. astronomical unit (the heliocentric distance at the time of the encounters). Because the temperature of sublimating ice is about 215 K, the location of sublimation must be below the surface.

The latest appearance of comet Halley may have led to the development of a technique to measure the temperature

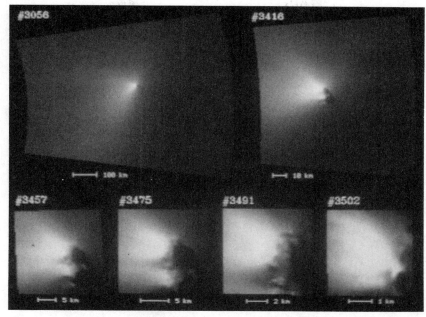

Figure 6.23 Sequence of images of the nucleus of Halley's comet taken by the Halley Multicolour Camera on ESA's *Giotto* spacecraft. (Courtesy of H. U. Keller, © Max-Planck-Institut für Aeronomie, Lindau/Harz, Germany)

of the interior ices before they are heated to the sublimation temperature. Infrared observations from the *Kuiper Airborne Observatory* at 2.65 micrometers (Figure D.6*e*) can determine the ratio of ortho water to para water. In ortho water, the nuclei of the H atoms spin in the same direction, while in para water the nuclei spin in opposite directions (see Figure 5.5). The ratio of these two species of water depends on temperature and changes very slowly. Current estimates indicate that the water ice in the interior of the nucleus can be sublimated and can flow through the crust into the coma without changing the ratio of ortho to para water. Thus the measurement of this ratio in coma gases tells us the ratio in the nuclear ice and therefore tells us the interior temperature. For comet Halley, the interior temperature was about 50 K. These measurements need to be carried out on other comets before a physical interpretation is attempted.

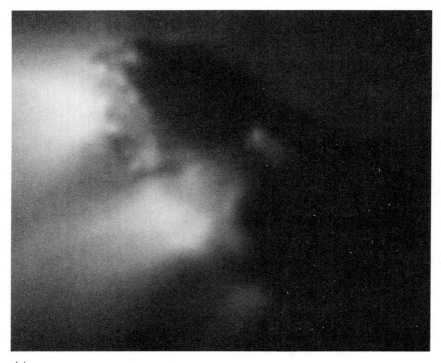

(a)

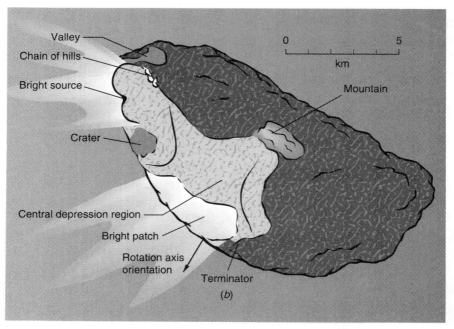

Valley

Chain of hills

Bright source

Crater

Central depression region

Bright patch

Rotation axis
orientation

Mountain

Terminator

0   5
km

(b)

(b)

The jets seen in the images occur only on the sunward side, and the emission is toward the sun. The jets are confined to an area equal to approximately 10 percent of the total area of the nucleus. They are visible by sunlight scattered off dust particles being dragged off the nucleus by the rapidly expanding gases. The gases expand laterally; the dust coasts outward and retains the signature of the original jet configuration.

The observation that the jets turn on rapidly at sunrise and turn off at sunset implies a thin dust crust over the jet areas and a rather thick crust elsewhere. Erosion of the "pits" from which the jets originate could produce Brownlee-type particles. The particles could originate from general erosion of the pit surface or at the pit edges.

An accurate description of the rotation of the nucleus is a major problem. Some constraints come from the orientation of the nucleus when the *Vega* and *Giotto* snapshots were obtained. Evidence for a 2.2-day (53-hour) rotation period was reported early in the apparition. This period appeared in variations in the Lyman-alpha brightness as it was observed by *Suisei* and was interpreted as pulsations or "breathing" of the nucleus. Other photometric work supported this period, and it was widely assumed to be correct during much of the apparition.

Observations from the *International Ultraviolet Explorer*, however, indicated a period of 7.4 days, as did observations from the ground. The problem of modeling the nucleus on the basis of all the available evidence has not yet been solved. When the comet is active, the turning on and off of the jets further complicates the problem.

The rotation of the highly asymmetrical nucleus is complex. It is almost certainly not as simple as the rotation of a symmetrical body such as a top. In that case, torques on the

Figure 6.24 (a) Composite of 60 images of the nucleus of comet Halley taken by *Giotto's* Halley Multicolour Camera. The sun is to the left and north is up. (Photo courtesy of Harold Reitsema, Ball Aerospace, © Max-Planck-Institut für Aeronomie) (b) Diagram to scale identifying the features in *a*. (Courtesy of Harold Reitsema, Ball Aerospace)

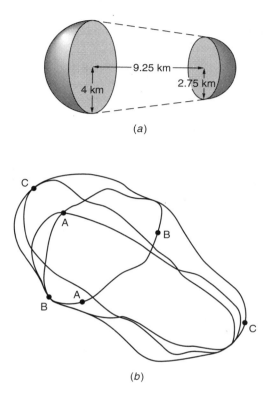

(a)

(b)

Figure 6.25 Possible representation of the nucleus of Halley's comet: (a) A simple axis-symmetric model with dimensions; (b) a more realistic model with unequal axes, which have the following approximate dimensions: A–A, 7.5 kilometers; B–B, 8.2 kilometers; and C–C, 16.0 kilometers. (After drawings by R. Z. Sagdeev et al., 1987)

body produce a precessional motion (which consists of a rotation of the axis of symmetry around a fixed axis) and a nutational motion (an up-and-down motion of the symmetry axis or rotation axis). If the torques exerted on the nucleus by the escaping gas and the torques produced by the gravitational attraction of the planets are negligible, the angular momentum vector must remain constant, but the rotation axis need not be fixed in the body. The rotation of the nucleus can be pictured as possibly occurring about all three axes simultaneously at different periods.

The most plausible model, according to Michael Belton and coworkers, has the long axis inclined to the angular momentum vector by 66 degrees. The nucleus wobbles in a manner that causes the long axis to rotate around the angular momentum vector with a period of 3.7 days. The nucleus also oscillates or rotates around the long axis with a period of 7.1 days. Additional observations of the comet as it recedes from the sun and the activity ceases may supply enough information to test this model.

Although our knowledge of comets has expanded tremendously, bear in mind that spacecraft have been sent to only two comets and data on only one nucleus have been obtained. There is no reason to assume that all comets are identical.

The Halley Findings

The studies of comets Giacobini-Zinner and Halley have provided us with an enormous quantity of new information about the two comets. The information has given us new insights into cometary physics; more important, it has confirmed many of the older insights we have gained over the years.

The in situ observations of comet Halley confirmed that Halley has a monolithic nucleus and in the process has strengthened our belief that all comets have monolithic nuclei. The compositional measurements have provided strong additional evidence in favor of Whipple's dirty-snowball model. The biggest surprise about the nucleus was its very low reflectivity.

The direct and indirect evidence for dust in both comets is about what we expected in terms of quantity and mechanical properties. The compositional studies of the dust revealed that there are several broad classes of grains, including particles composed of the lighter elements carbon (C), hydrogen (H), oxygen (O), and nitrogen (N)—the CHON particles.

The studies of the magnetic fields and the plasma phenomena in both comets verified the basic features of a model put forward by Alfvén. The direct measurements of the composition of the neutral and ionic chemical species in Halley have verified the presence of most of the species observed before the encounter and have provided evidence of new species as well. One of the interesting new species was polymerized formaldehyde.

The missions did not answer all the outstanding questions in cometary science, so cometary researchers have a bright future to anticipate. (Chapter 9 discusses what may come next.)

The Death of a Comet

What eventually happens to the cometary material and the main cometary body? The gas and small dust particles are swept out of the solar system by the pressure of solar radiation and the solar wind. The intermediate-size dust particles are not so strongly influenced by radiation pressure and continue to orbit the sun. Reflected sunlight from these particles is seen in the night sky as the faint glow called the zodiacal light.

The largest dust particles stay in nearly the same orbit as the comet (see Plate 13), and as the comet becomes extinct by sublimating all its ices, more and more particles are produced. They may be fluffy, like the Brownlee particles, or denser, depending on the details of the interior. Gravitational perturbations distribute the material along the orbit. The particles produce meteor showers when they enter the earth's upper atmosphere.

We can crudely estimate the time scale for the ultimate demise of a comet by noting that comets lose roughly 0.1 to 1 percent of their mass each time they appear. This means that they should last roughly 100 to 1000 apparitions. Given the vast phenomena produced, the loss of material at each apparition is surprisingly modest. A hypothetical comet with a radius of 5 kilometers that is sublimating material uniformly over its entire surface would lose only the top 2 meters at each apparition.

Eventually all the ices sublimate and the comet is extinct. Depending on the interior details, the comet could completely disintegrate into smaller particles, or one or more extinct large bodies could remain. These bodies would resemble asteroids.

A Physical Understanding of Comets

Cometary scientists have attempted to boil down all the knowledge they have gained into a set of simple summary diagrams, to put everything into context. Figures 6.26 and 6.27 are results of that attempt. Figure 6.26 is a schematic illustration of the principal features observed in comets; Figure 6.27 is a grand summary of the physical processes believed to be important in comets.

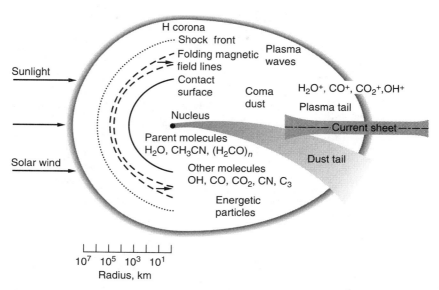

Figure 6.26 Summary of cometary features and phenomena. Note the logarithmic scales.

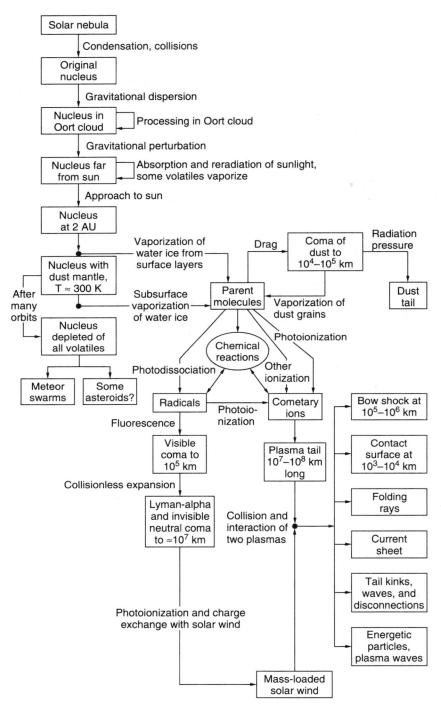

Figure 6.27 Grand summary of physical processes in comets.

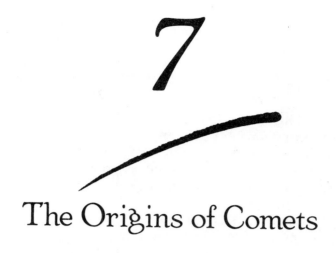

The Origins of Comets

Our knowledge of the origins of comets is much less secure than our knowledge of physical models. The raw materials for our detective work are the nuclear properties, especially the chemical composition of the nucleus, and the orbital properties of the comets observed in the inner solar system. From this information and the application of physical principles we should be able to deduce the region and physical conditions of formation, the region of storage if necessary, and the orbital evolution of comets.

Our discussion concentrates almost entirely on the current paradigm of the formation and evolution of the solar system. Comets are viewed as natural products of the processes that formed the solar system, the same processes that created the sun, the planets, and the asteroids.

The Birth of a Comet

The sun was formed by the same sort of processes that formed the stars throughout our galaxy. A volume of the interstellar gas collapses, ultimately to form stars and a planetary system. Scattered here and there in our galaxy are dense clouds of gas and dust (nebulas) that need just a small increase in density to begin to collapse under their own gravitational forces. The appropriate nudge could occur in a number of ways. For instance, the spiral arms in our galaxy are actually spiral-shaped density waves that move slowly through the interstellar material. It is an observed fact that the very youngest stars in the galaxy are situated at the present locations of those arms, an indication that the passage of the density enhancement has stimulated the formation of the stars.

We believe that a cloud collapses in the form of a central body and a disk. As the collapse proceeds, fluffy layers of ice build up on the dust grains. The resulting ice balls become more and more likely to collide as the cloud shrinks. When they do collide, they stick together and form increasingly larger bodies. If we could watch the process as it proceeds, we would see the cloud develop into a central mass that would eventually become the sun and a large disk-shaped collection of solid bodies that would become the planets, comets, asteroids, and other interplanetary matter. The dynamics of the formation of the disk require the initial cloud to have some angular momentum, so that the forming system rotates. It is this angular momentum that ultimately leads to the common direction of orbital motion of the planets and their moons.

Dust grains in the disk are believed to exhibit a core-mantle structure—that is, a refractory core and an icy mantle. In the inner part of the solar system, the icy mantle is sublimated away by the heat of the forming sun, and agglomerations form planetesimals and, ultimately, planets. The Jovian planets presumably were massive enough to capture and retain an envelope of approximately solar composition. Certainly there are many details yet to be understood in this picture.

Far away from the sun, the icy dust grains could serve as condensation centers for the formation of icy bodies. These bodies could grow by additional condensation and agglomeration to cometesimals, the building blocks of comets and the outer planets. Just where is this comet factory?

The distance is estimated by considering the maximum temperature at which key volatiles (substances that evaporate at low temperatures) may be condensed or preserved. Temperatures of roughly 100 K are needed to form clathrate hydrates. If diatomic sulfur (S_2) is present at or formed during the condensation process, temperatures of 15 K or less are required. In addition, the particles must coagulate at low speeds, and the effective conductivity of the nucleus must be very low to insulate the inner layers. To have diatomic sulfur in comets, the factory must be located at least 1000 astronomical units from the sun if it is bathed by almost full sunlight. If the nebula is dense enough to shield the comet factory from the sun, the restriction on distance is removed. Whether this region is otherwise plausible—that is, of sufficient density—remains to be seen. Radiation pressure on regions of enhanced density could be important in compressing interstellar grains in this region. Once the comet is well into the inner Oort cloud or the classical Oort cloud (discussed in the next section), it will usually be at approximately the temperature of a black body in the interstellar medium, 10 K or less.

It is widely believed that after a comet is formed and before it is exposed to the environmental rigors of the inner solar system, it is in cold storage, undergoing little or no evolution. It could collide with other comets, however, and its surfaces could be subjected to cosmic rays and radiation from passing stars and supernovas. If bombardment by cosmic rays produced diatomic sulfur during this phase, for example, the comet factory could be nearer the sun (possibly as close as the region of Uranus or Neptune—Neptune is at 30 AU), or the requirement to shield the comet-forming region from the sun could be dropped. If comets were formed as close to the sun as the region of Jupiter, most of them would have been strongly perturbed by Jupiter and thrown out of the solar system. Thus the comet factory could have been as close to the sun as the Uranus/Neptune region and as far out as beyond 1000 astronomical units.

The Oort Cloud Revisited

Jan Oort proposed in 1950 that the Oort cloud was the storage area for essentially the entire comet population. After formation in the comet factory region, the comets were perturbed by the planets. The comets slowly diffused into the distant Oort cloud, where their orbits were then somewhat circularized by stellar perturbation. Stars passing at random occasionally perturbed comets back into the planetary region.

The concept of the Oort cloud evolved considerably with the realization that encounters with passing stars, and especially giant molecular clouds, would completely dissipate the classical Oort cloud over the lifetime of the solar system. The theory that the cloud is the source of new comets requires a source of replenishment of the cloud itself. Thus the concept of the massive *inner Oort cloud* was born. The same giant molecular clouds that strip away the more loosely bound comets in the outer cloud also pump comets from the inner to the outer cloud. The inner cloud was thought to extend from about 35 to 20,000 astronomical units, and these dimensions are entirely compatible with the discussion on the region of formation. The inner cloud probably contains 10^{13} to 10^{14} comets with a combined mass of 10^2 earth masses. The outer (classical) cloud, extending from 10^4 to 10^5 astronomical units, contains 10^{12} to 10^{13} comets and a total mass of about 10 earth masses. The two clouds should be considered as a continuum; in other words, there is no sharp boundary between them. The shape is a flat disk at the inner boundary, expanding to nearly spherical for the classical cloud. We must stress that the inner cloud remains a theoretical construct, supported by a large body of indirect evidence but no direct evidence.

The Evolution of an Orbit

After a comet forms in the outer reaches of the solar system, it spends most of its lifetime in the inner Oort cloud. Eventually its orbit reaches the outer Oort cloud, where it is strongly perturbed by passing giant molecular clouds and stars. Most of the time it is lost to the solar system, but sometimes it is sent into the region of the planets, where we see the display of the so-called cometary phenomena. These are the fresh comets with periods of approximately 5 to 10 million years from the outer reaches of the Oort cloud—that is, aphelion distances of 60,000 to 90,000 astronomical units. The inclinations of these comets' orbits to the ecliptic have essentially a random distribution.

Each time a comet passes through the planetary region it suffers a change in energy because of the gravitational effects of the planets. The energy is often expressed in terms of the reciprocal of the semimajor axis of the orbit, $1/a$ (a in AU). The average change in energy in units of $1/a$ is 0.0005, and the change can be positive or negative. A comet from the Oort cloud with an aphelion of 100,000 astronomical units has a q value of 50,000 astronomical units and a $1/a$ value of 0.00002. Thus a positive average change would bind this comet more strongly to the sun, while a negative average change would produce a negative total energy, and the comet would be ejected from the solar system.

Now, where do the short-period comets come from? These are the comets that have periods of less than 200 years, are generally prograde, and have orbits of low inclination. For some time the explanation that a comet could be captured in a short-period orbit by a single close encounter with a planet has been considered inadequate. Rather, the creation of the short-period comets from the population of long-period comets was considered to be the cumulative result of many passes through the planetary system; each pass would produce the small change in energy just described. Certainly, some short-period comets can be made in this way.

There is a problem of efficiency, however, and the solution may involve the inner Oort cloud. Planetary perturbation acting on a population of long-period comets of all inclinations produces a population of short-period comets that also are at all inclinations. This problem vanishes if the source is a belt of comets in the plane of the ecliptic near and beyond the orbit of Neptune. Recall that the inner boundary of the comet storage region lies at about 35 astronomical units and that the comets there are confined to a flat disk.

Two difficulties remain with this otherwise promising approach. First, some source of gravitational perturbations—perhaps Neptune, perhaps very large comets in the cloud, perhaps the long-searched-for tenth planet—must be hypothesized so that the effects of the perturbations are great enough to start the required number of comets moving inward. Second, a close-in belt population of comets produces gravitational effects, depending on the total mass of comets and the distance of the belt. Studies of the orbit of Halley's comet rule out some combinations, but others are possible, such as a belt at 100 astronomical units with a total of 1 earth mass. Again, these are exciting developments, and some clarification is expected over the next few years.

Meanwhile, this is the picture as we see it now. The innermost part of the comet storage area, called the Kuiper-Duncan disk, lies between 35 and 1000 astronomical units from the sun and contains some 10^7 to 10^9 comets. This disk, originally proposed by Gerald Kuiper years ago and more recently discussed by Martin Duncan, includes the region where comets are formed. The inner Oort cloud, also disk-shaped, lies between 1000 and 20,000 astronomical units from the sun and contains some 10^{13} to 10^{14} comets. The outer Oort cloud lies between 20,000 and 100,000 astronomical units from the sun and contains some 10^{12} to 10^{13} comets. These demarcations are partly for convenience; as we have said, there are no sharp boundaries. All estimates of numbers are somewhat suspect.

Comets are usually lost by being ejected from the solar system, and this could have occurred at any time during the comet's lifetime of 4.5×10^9 years. Some comets that are permanently trapped in the inner solar system are lost violently by collision with the sun, planets, or satellites. The rest simply lose all their volatile materials and cease to be comets. Then they become asteroids or meteor swarms. A grand summary of the origins and evolution of comets, a companion to Figure 6.27, is seen in Figure 7.1.

Origins: An Assessment

Our discussion of origins, to put the matter politely, has been equivocal in some places. The theory of the origin of comets is a subset of the theory of the origin of the solar system, a

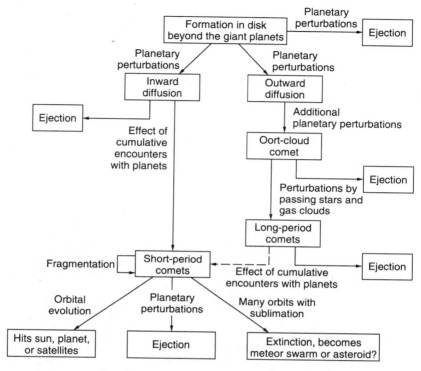

Figure 7.1 Grand summary of orbital evolution and terminal states. Obviously, not all possibilities are included. The solid line to the short-period comets represents the primary path, and the dashed line represents a secondary path. Note that ejection from the solar system is a comet's most likely fate.

massive problem. Unfortunately, the origin of comets has often been overlooked in the broader discussion of origins. This discussion must begin from an unknown initial situation, must not violate the laws of physics, and must reproduce essentially all known details of the solar system! In this milieu, ideas can be a very perishable commodity. Nevertheless, ideas serve the cause of scientific advancement in their day, even if they are eventually found to be wrong. In the next few years more ideas will be laid to rest and new ones will appear. This is the way science advances.

Some indication that we are on the right track comes from studies of main sequence stars by the *Infrared Astronomy Satellite (IRAS)*. Main sequence stars are stars in the chief phase of their life, when they convert the hydrogen in their cores to helium to produce the energy they emit. Many such stars were observed to have low-temperature dust shells or disks around them extending to distances that would be in our inner Oort cloud and show signs of cleared zones closer to the stars. One such star, Beta Pictoris, is shown in Figure 7.2. It is widely considered plausible that the evolution of a solar system by agglomeration into solid bodies and dispersal of debris is faster near the stars. Presumably the disks observed by *IRAS* are characteristic of young systems. Older systems would surely disperse more of the original material and make the disks harder to detect. The disk-shaped comet population that constitutes the inner part of the inner Oort cloud, however, might be de-

Figure 7.2 An image of Beta Pictoris taken in red light on March 30, 1987, at the Cerro-Tololo Interamerican Observatory, showing dust disks extending to distances of approximately 900 astronomical units. (Courtesy of S. M. Larson and S. Tapia, Lunar and Planetary Laboratory, University of Arizona)

tectable even for an older system. Thus we can check some of our ideas, particularly those related to the early conditions in the formation process, by observing other star systems in the infrared.

Comets and Other Stars

Why Don't We See Interstellar Comets?

Robert Chapman and a colleague have estimated the number of interstellar comets that should move through the inner solar system and be observable if current ideas are correct. The theories of the origins of comets postulate that they are initially formed in the vicinity of the outer planets, then slowly diffuse out into the Oort cloud. This diffusion process is very inefficient, and a significant fraction of the comets are completely lost to the solar system. Current calculations indicate that somewhere between 30 and 100 comets are lost for every one that eventually winds up in the Oort cloud. If we accept the estimate that there are 2×10^{12} comets in the Oort cloud today, then as many as 10^{14} comets may have been lost to interstellar space.

If our solar system is typical of planetary systems around other stars, then we can estimate that there are roughly 10^{14} interstellar comets for every star. Given the observed density of stars in the solar neighborhood, we estimate that there are 10^{13} comets in each cubic parsec (30.9 trillion cubic kilometers) of space near the sun. On the basis of that number, 0.6 extrasolar comets should pass within 2 astronomical units of the sun each year.

The comet researcher Edgar Everhart has studied the probability of detecting long-period comets. On the basis of his work, we estimate that the probability of detecting an extrasolar comet is 0.07. This line of reasoning leads us to conclude that we should have seen six extrasolar comets in the last 150 years.

How could we tell an extrasolar comet from a solar-system comet? An extrasolar comet would move through the solar system on an orbit that is clearly hyperbolic. In the last 150 years many comets whose eccentricities appear to be slightly hyperbolic have been sighted. Typically, however, we calculate a comet's orbit in relation to the center of the sun. If instead we refine the calculation and determine the orbit in relation to the center of mass of the solar system, many of the slightly hyperbolic orbits become elliptical, an indication that the comets are bound to the solar system. The remaining hyperbolic orbits are easily explained by observational error or by perturbations by the giant planets. We have never seen a comet moving in an obviously hyperbolic orbit.

Something is amiss. Clearly we must look more carefully at our theories to see if the problem lies with one or more of our hypotheses. Perhaps solar systems like ours are the exception rather than the rule. Perhaps our ideas about how the Oort cloud formed are not correct. Either of these possibilities could lead us to conclude that there are fewer interstellar comets in the vicinity of the sun than we estimate. More research is certainly needed along these lines. One bit of evidence that would really be important would be the direct detection of Oort clouds around other stars. Astronomers are beginning to look for other Oort clouds.

Do Other Stars Have Oort Clouds?

Direct detection of Oort clouds around other stars would be an exciting discovery. It would establish the generality of the phenomenon, enable us to refine estimates of the number of interstellar comets, and provide an argument for the formation of planetary systems around other stars.

So far the search for the infrared emission that would indicate an Oort cloud around a nearby star have been unsuccessful. There is another interesting possibility, though. When the sun finishes its life on the main sequence, it will evolve and become a much more luminous giant star. Stars like the sun or cooler that evolve to the giant or supergiant phase can have typical luminosities in the range 6×10^3 to 6×10^4 suns for periods of one million to several hundred million years. This sustained, very large increase in a star's luminosity can produce dramatic effects in the surrounding cloud of comets.

This process can be modeled fairly simply, and the result is that water ices sublimate rapidly at distances up to about 500 astronomical units, in contrast to a distance of about 3 astronomical units for the present sun. The sublimation can occur in comets on essentially circular orbits within this distance; these are the comets of the Kuiper-Duncan disk. Sublimation can also occur in Oort-cloud comets on eccentric orbits that dip into this region. Figure 7.3 shows the rate at which mass is lost and the survival time of a cometary disk around a star 6000 times as luminous as the sun.

Curiously, many giant and supergiant stars are surrounded by extended regions where water and hydroxyl are emitted, and red giants exhibit both water and ice grains in

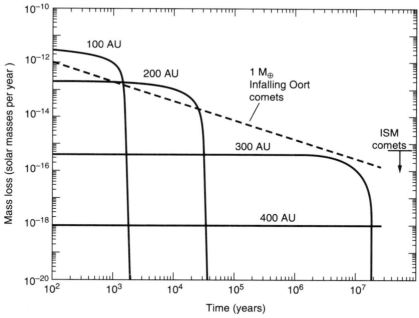

Figure 7.3 The destruction of comets around a star with luminosity 6×10^3 times that of the sun. For the curve labeled "100 AU," mass is lost quickly at first, but the rate soon drops after all comets within 100 astronomical units are sublimated. The process is repeated for comets farther from the star. The contribution from Oort-cloud comets and interstellar (ISM) comets is also shown. (After S. A. Stern, J. M. Shull, and J. C. Brandt, University of Colorado)

their outflows at distances of several hundred astronomical units. The observed range of rates at which some stars lose mass is compatible with the calculations for the comet-cloud sublimation model. But sublimating comets are not the only proposed explanation for the water in stellar outflows; the competing mechanism is photochemistry.

The comet sublimation model can be tested in several ways. The rate at which mass is lost decreases with time. Because the sublimation operates on a disk-shaped source, the emission should have an annular, or ringlike, shape (in contrast to the more spherical shape expected from stellar outflows). Finally, individual sublimating bodies on eccentric orbits may explain the irregular and variable blobs observed in the emitted water and hydroxyl.

The sublimating comet idea has certainly not been proved; it needs to be explored and tested. Establishing the existence of extrasolar Oort clouds would be an exciting event and would help us understand the probability of the formation of comets and planetary systems in the galaxy.

Comets and the Solar System

One theoretical scenario of comet formation—a series of ever-larger solid bodies forming in a sequence (condensing gas, dust, cometesimals or planetesimals)—encourages us to imagine many roles for comets in the solar system, both in the past and at present. A survey of comets' roles in the solar system must begin with a description of the first historical connection between comets and other phenomena of the solar system.

Comets and Meteor Streams

Over a century ago Giovanni Schiaparelli established that the orbits of some meteor streams were quite similar to those of some specific comets. The visible meteor phenomenon is produced when small solid particles enter the earth's atmosphere and burn up by friction at altitudes of about 100 kilometers (Figure 8.1). Meteor streams consist of particles, all moving in essentially the same orbit, entering the earth's atmosphere. Because their paths are parallel, the meteors appear to emanate from a fixed point on the celestial sphere, called the radiant. Orbits of the *meteoroids*—the small bodies in space that produce the meteors—can be calculated, and the similarity between the orbits of specific comets and the orbits of meteoroids has been established.

The connection implies that some solid particles produced by an active comet or left behind after a comet becomes inactive are large enough to stay in an orbit similar to the comet's. The production of a variety of solid particles as a result of erosion or evolution of the crust layer on a nucleus is a reasonable, simple process. The larger particles produce meteor showers; the fates of the smaller ones depend on the effects of solar radiation pressure and the following effect. When solar photons reflect from minute orbiting dust particles, the photon imparts momentum to the particle. Because of the relative motions of the photon and the particle, some of the momentum is imparted in a way that slows the particle in its orbit, and the deceleration causes the particle to spiral in toward the sun. The effect, known as the *Poynting-Robertson effect,* sweeps the dust particles into the sun.

Cometary Particles

We may think of dust particles from comets as falling into three categories: (1) the small particles, which experience strong solar radiation pressure and are blown out of the solar system, form the comet's visible dust tails; (2) the intermediate-size particles, which experience some force from radiation pressure but not enough to blow them out of the solar system, go into

Figure 8.1 A meteor observed in the constellation Cygnus. Most meteors are believed to be pieces of solid material from comets that are burned up by friction as they enter the earth's atmosphere. (Yerkes Observatory photograph)

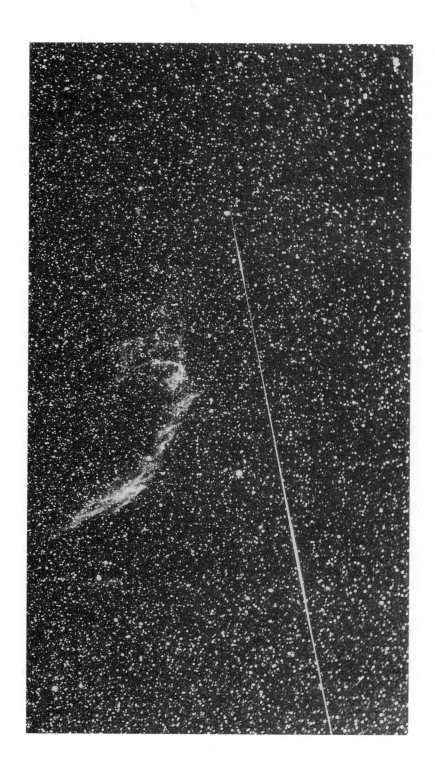

fairly long-lived orbits around the sun, produce the zodiacal light by reflected sunlight, and are collected when they enter the earth's upper atmosphere (these particles experience the Poynting-Robertson effect described above); and (3) the large particles, which experience negligible radiation pressure, continue in the same orbit as the comet. Gravitational perturbations distribute the particles along the orbit, and meteor streams are produced. Disrupted swarms can be responsible for some sporadic meteors. Bear in mind that gravitational effects can deflect the orbit of the meteor swarm from that of the original comet.

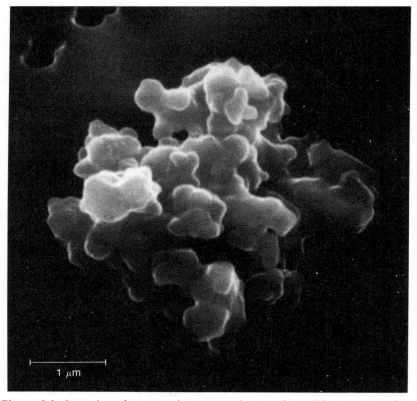

Figure 8.2 Scanning electron microscope picture of possible cometary dust particles (Brownlee particles) collected in the earth's atmosphere. (Courtesy of Donald Brownlee, University of Washington)

The collections of dust particles by aircraft flying in the earth's upper atmosphere have supplied samples of what is probably cometary dust for examination. Figure 8.2 shows some of these Brownlee particles. These rather large, complex particles, of low density and with voids, could be described as fluffy.

The nature of these particles is consistent with our expectations that they originate as products of crust erosion. The polarization measurements of cometary comas show behavior characteristic of light that has experienced multiple reflections in a complex structure such as that of the Brownlee particles. Finally, the chemical composition of these particles is unlike that of any known terrestrial particles. If these particles originated on earth, their chemical composition would readily identify them. Because these particles are unlike others that we know of, scientists conclude that they are of cometary origin.

The Zodiacal Light

Another dramatic reminder of the pervasiveness of cometary dust in the solar system is the zodiacal light (Figure 8.3). Dust

Figure 8.3 The zodiacal light as seen from Mount Chacaltaya, Bolivia. (Courtesy of D. E. Blackwell and M. F. Ingham)

particles in orbit around the sun are constantly removed (by the Poynting-Robertson effect), and a source of new particles is therefore required. Dust from comets replenishes the zodiacal light cloud. Comet Encke, which has passed through the inner solar system many times, is clearly one major contributor; the fraction of particles supplied by other comets and other sources is uncertain.

The *IRAS* observations of comet Tempel 2 (Figure 8.4 and Plate 13) in 1983 demonstrate that comets are a prolific source

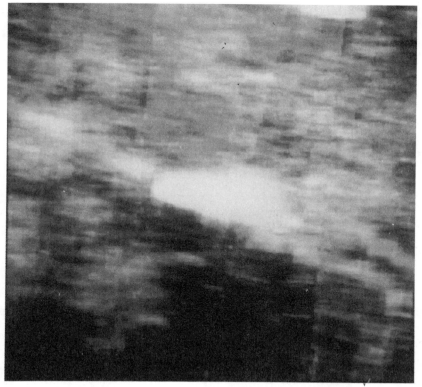

Figure 8.4 *Infrared Astronomical Satellite (IRAS)* observations of comet Tempel 2 and its dust trail on September 6, 1983. The comet (fishlike object) consists of particles tens of microns in radius. The trail (the streamlike object passing through the comet and sloping down to the right) is composed of particles at least a centimeter in radius ahead of the comet (left) and at least a millimeter in radius behind the comet (right). (Courtesy of Mark Sykes, University of Arizona)

of dust. Indeed, the trail of large particles from comet Tempel 2 appears to be a meteor stream in the making. The infrared observations clearly show the large dust efflux from comets, particularly the larger particles.

Figure 8.5 is an interesting illustration of cometary dust in the solar system. A thin, spikelike antitail is seen in this photograph of comet Arend-Roland made on April 27, 1957, but no antitail is visible in photographs made a few days earlier or later. Researchers wondered: Could there be a set of conditions in which some dust particles are attracted toward the sun? The mystery of the antitail was cleared up when researchers realized that the earth crossed the plane of the comet's orbit on April 25. The antitail was not a tail at all; it was the layer of dust that outlines the comet's orbit scattering sunlight. Normally this thin layer of dust cannot be seen; but when it was viewed edge on, during the time the earth was passing through the plane of the orbit, it briefly became visible.

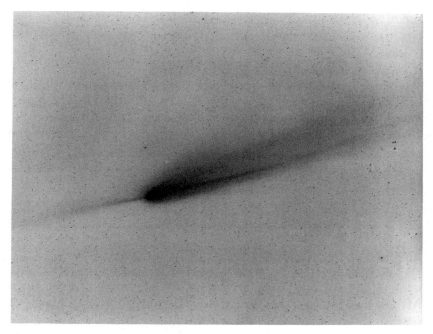

Figure 8.5 Comet Arend-Roland on April 27, 1957. The antitail is an apparently sunward spike composed of dust particles. (Palomar Observatory photograph)

Comets and Asteroids

Some observers have wondered whether some asteroids might be extinct short-period comets—that is, comets that have evolved into low-activity objects. Two distinct lines of evidence bear on this question. The first, dynamical evidence, comes from investigation of orbital evolution; the second, physical evidence, comes from comparisons of the observed morphological properties of comets and asteroids.

Some of the dynamical evidence is frankly circumstantial. In Figure 8.6 we plot the eccentricities of the orbits of comets and asteroids on the abscissa and the semimajor axis of the orbit on the ordinate. Note that the asteroidal orbits tend to lie in the lower part of the diagram and the cometary orbits at the upper right. Of most interest to us are the exceptions to the general rule.

One of the obvious exceptions is the asteroid Hidalgo, which lies well within the cometary region of the diagram. Astronomers have studied the motion of Hidalgo and see evidence of weak nongravitational forces. Hidalgo is probably an extinct comet.

Within the asteroid region is a group of comets, including P/Oterma and P/Gehrels 3, that have orbits very similar to those of Hilda asteroids. These comets are sometimes called quasi-Hilda comets. Comet P/Gehrels 3 is interesting because of its strong interactions with Jupiter. It has been observed to be temporarily captured by Jupiter, to make three revolutions around the planet, and then to move away again. Astronomers believe the quasi-Hilda comets are evolving toward asteroids.

Comet Encke moves in an asteroidlike orbit. There is evidence that the comet's gas output is slowly decreasing; it may be approaching extinction. Once it is extinct, it will be indistinguishable from an asteroid.

A very interesting group of objects consists of the earth-crossing asteroids. There are actually three recognized groups of those asteroids. The Aten asteroids have semimajor axes of less than 1.0 astronomical unit and aphelion distances that are larger than the earth's perihelion distance (0.983 AU). Thus an Aten asteroid's orbit overlaps the earth's orbit when the asteroid is near aphelion. The Apollo asteroids are the oppo-

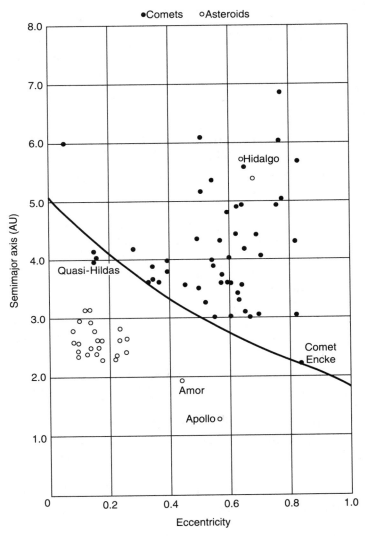

Figure 8.6 Orbits of asteroids on a plot of semimajor axis versus eccentricity. (Adapted from a figure by L. Kresák)

site case: they have semimajor axes greater than 1.0 astronomical unit, and their perihelion distances are less than the earth's aphelion distance (1.017 AU). The orbits of the Apollo asteroids overlap the earth's orbit near their perihelion. The Amor asteroids have perihelion distances between 1.017 and 1.3 astronomical units and may approach close to the earth's orbit, but do not overlap it. Today we know of only 3 Aten asteroids, 23 Apollo asteroids, and 20 Amor asteroids, but this known group leads researchers to believe that there are probably 1000 or more fainter earth-crossers bright enough to be found someday.

Earth-crossing asteroids have very short lifetimes (about 30 million years) in comparison with the age of the solar system. They can be destroyed by collisions with the earth or thrown from the solar system by close encounters with the earth. The probability of such an event is relatively high, and the population of earth-crossers will disappear if some process does not replenish them. Astronomers believe that short-period comets are a likely source of new earth-crossing asteroids, in part because they cannot identify any other suitable source. As we said at the outset, the dynamical evidence that some comets might evolve into asteroids is circumstantial.

The other line of evidence comes from the physical observations of low-activity comets and asteroids. There is a problem in interpreting these data. Note that the observations are of the bare cometary nucleus, not of a faint coma. Two comets (P/Arend-Rigaux and P/Schwassmann-Wachmann I) have been observed at times when it is likely we were seeing the nucleus. Then their optical spectra resembled the optical spectrum of asteroid Hidalgo.

Researchers have concluded that we have no direct evidence that an extinct cometary nucleus has been observed. It is clear to these researchers that scientists must continue to study low-activity comets and faint asteroids to build up the body of data that will help us answer the interesting question of the relationship between comets and asteroids.

When Comets Crash to Earth

We have seen that comets can come very close to the sun, and even hit it. There is evidence that comets have collided with both the moon and the earth as well.

Let us look at the earth first. We know of more than 100 fossil craters on the earth, with sizes ranging from a kilometer or so to over 100 kilometers. Most of these craters are relatively recent, as geological time goes, because old craters would have been completely obliterated by the weathering process. One of the larger craters is a 70-kilometer circular lake in Quebec, which is estimated to be 200 million years old. Impacts that produce craters that size and larger probably occur once every 30 million years or so. It is difficult to be sure which of the known craters result from collisions with comets and which from asteroids.

Eugene and Carolyn Shoemaker have made it their life's work to study asteroids and to look for evidence of collisions between large bodies and the earth. They estimate that more than 1000 kilometer-sized comets and asteroids are moving in orbits that could cause them to collide with earth sometime in the future. About one-third of this population is made up of comets, so we might estimate that roughly one-third of the giant craters can be attributed to comets.

Asteroid 1989FC, which may be a half-kilometer in diameter, was discovered on March 31, 1989. A subsequent series of observations permitted a preliminary orbit to be calculated, and it indicated that the asteroid had actually crossed the earth's orbit on March 23 and missed us by about 700,000 kilometers. That distance, nearly twice the moon's distance from earth, may seem like a comfortable miss until you realize that this is the distance the earth moves along its orbit in just six hours. If the asteroid's arrival time had varied by six hours, we could have had a natural disaster of catastrophic dimensions.

Encounters with smaller bodies are more frequent. The Meteor Crater near Winslow, Arizona, is the result of the impact of a mostly iron body only a few tens of meters in diameter, which was traveling at over 40 kilometers per second when it hit. The crater is roughly 1.2 kilometers across and over 100 meters deep (Figure 8.7). It has been estimated that the impact took place between 20,000 and 40,000 years ago.

There is also evidence that comets have crashed on the moon. There are some unusual bright swirled markings on the moon, mostly on the side facing away from the earth. One of the most prominent of the markings is in the vicinity of the crater Reiner Gamma, which we can see on the edge of the moon's near side. Researchers have studied these swirls through sensitive telescopes on Mauna Kea in Hawaii and find that their

Figure 8.7 The Meteor Crater near Winslow, Arizona. (Yerkes Observatory)

infrared reflectivity does not match that of any other known lunar feature. The researchers suggest that cometary gas and dust hitting the moon would stir up the lunar soil. We believe that the lunar surface has been darkened by exposure to billions of years of solar wind. The deeper material, however, would not have been exposed and would be light in color. If a comet had hit relatively recently (say, only 100 million years ago), the exposed material would still be fairly light-colored. It is quite possible that this is the explanation of the swirls.

We know that the tail of comet Halley may have hit the earth in 1910. On May 18, 1910, the comet passed directly between the earth and the sun. The event was interesting because the tail—which points away from the sun—should have come very close to earth on May 19. Either it just missed us or the tail gases simply produced no observable effects.

Denser cometary gas and dust might affect the lunar surface, but it could never reach the surface of our planet through its protective atmosphere. The density of the cometary gases is so low that they are stopped high in the earth's atmosphere. Even if they could eventually reach the ground, they would be so diluted that the effect would be negligible—fortunately, because some of the constituents of cometary gas, such as cyanogen (CN) and cyanogen compounds (hydrogen cyanide [HCN] and methyl cyanide [CH₃CN]), are deadly poisons. The gases released in the Bhopal tragedy in India were cyanogen compounds that probably were broken down to the highly poisonous hydrogen cyanide. In 1910 many people feared that the comet gases might poison the atmosphere, because the media did not fully explain the amounts of the gases and the protective effect of our atmosphere.

Tunguska

What if the nucleus itself hit the earth? In the summer of 1908 a huge explosion, called the Tunguska event, took place in central Siberia. An earthquake was registered at a seismic station at Irkutsk, almost 900 kilometers from the site of the explosion. Meteorological stations around the world measured an air blast that dwarfed any known volcanic eruption.

The area is sparsely inhabited, but the local peasants reported seeing a bright object in the sky, moving rapidly earthward. As the object "approached the ground it seemed to be pulversized, and in its place a huge cloud of black smoke formed and a loud crash, not like thunder, but as if from the fall of large stones or from gun-fire, was heard."[1]

Looking at the evidence available, scientists assumed that the event was a giant meteorite striking the earth. In 1921, 13 years after the event and 4 years after the revolutions of 1917, L. A. Kulik of the Russian Academy of Sciences went looking for the fall site. Kulik never got there himself in 1921, but he was able to talk to several eyewitnesses. Their stories, plus the seismic data, helped pinpoint the actual location of the site.

Kulik finally made it to the site in 1927, after a valiant 75-mile trek through deep snow and dense forest. He found a shallow but extensive depression filled with craters 10 to 50 meters across. Within the impact area the forest was com-

---

1. E. L. Krinov, *Giant Meteorites* (New York: Permagon, 1966), p. 128.

pletely flattened. The trees, stripped of limbs and bark and stacked like so many pickup sticks, all pointed away from the central point (Figure 8.8). The local residents claimed that a sizable herd of reindeer in the area simply vanished. The biggest surprise was that there were no meteorites in evidence at the site. Whatever hit was completely destroyed.

In 1962 another scientific expedition went to the area and used a helicopter in their research. This group of scientists gathered soil samples in search of cosmic dust particles. They found a narrow area almost 250 kilometers long that contained unusual amounts of cosmic dust. The Czech astronomer Lubor Kresák collected all the sightings of the object and tried to reconstruct its orbit. The result was similar to the orbit of comet Encke. Kresák suggested that a large chunk broke off from the nucleus of Encke's comet and then hit the earth. Why

Figure 8.8 The Tungska event. These trees were blown down approximately 8 kilometers from ground zero. (Sovfoto)

didn't we see the comet? According to Kresák's calculations, it would have approached the earth from the direction of the sun and would have been lost in the glare.

If Kresák's idea is correct, the chunk would have been moving at 150,000 kilometers per hour when it hit the earth's atmosphere. At this great speed, each gram of the chunk would have had as much energy of motion (kinetic energy) as the chemical energy in a gram of TNT. Given the wide extent of the destruction, the chunk was probably about 50,000 tons in mass (equivalent in destructive potential or force to a 50-kiloton nuclear weapon). If this was an ice ball, it was probably about 40 meters in diameter. The object, ice ball or otherwise, probably never reached the ground, but broke up at an altitude of roughly 10,000 feet because of atmospheric friction. The destruction was caused by a pressure wave in the atmosphere and a few surviving small pieces of the chunk.

The cometary explanation of the Tunguska event has been the standard one for years, but recent study shows that it is probably not correct. The comet scientist Zedenek Sekanina only a few years ago asked a crucial question: Could a comet nucleus as we now picture it—a loosely compacted dirty snowball—survive passage through the atmosphere down to approximately 10,000 feet? The answer depends on specific details, such as the size of the comet, the slant of its path through the atmosphere, and whether it is composed of cometesimal building blocks. Larger comets might survive to lower altitudes or even reach the surface by sublimating their outer layers. Larger comets might break into building blocks roughly estimated to be 0.5 kilometer in diameter. And the slant path determines the length of time the nucleus is heated by atmospheric friction.

The Tunguska body was small and had a large slant path. So no, a small comet nucleus cannot survive passage through the atmosphere down to 10,000 feet. Because of the expected tensile strength of a typical cometary nucleus, it would break up high in the atmosphere. We now think that the body was probably a fragment of an asteroid rather than a comet. If the similarity between its orbit and that of comet Encke is significant, connection with a comet cannot be entirely ruled out. The object conceivably could have been an inert, rocky piece from a comet. We are now in a semantic problem, but the object was unlikely to have been cometary in the dirty-snow-ball sense. Objects that originated as asteroids, not as comets, are the likely causes of such craters as those in Arizona and Siberia.

The revised view of the Tunguska event does not rule out important encounters between comets and the earth and other planets. There are many comets out there, and collisions must occur. To understand the importance of such impacts, we need to determine how often they occur for a comet of a given size. The frequency depends on the number of comets in different size ranges in the inner solar system. The comets we observe typically range in size from 1 to 10 kilometers, and the larger ones are extremely rare. But could there be large numbers of small comets that would have a higher collision rate and thereby produce significant terrestrial effects?

Large numbers of small comets could be invoked to explain certain phenomena of the upper atmosphere and observations of hydrogen in interplanetary space. The collision rates needed are probably too high, but even lower rates can have interesting consequences. If cometesimals 15 meters in diameter, one-half of which is water ice by volume, struck the earth only once a minute over 4.5×10^9 years, they could supply the entire water content of the oceans and atmosphere. The impact rate could be even lower now, because the density of cometesimals is believed to have been generally much higher in the past and to have been enhanced during cometesimal showers triggered by the close passage of stars and interstellar clouds through the Oort cloud. This line of thinking is controversial but it is clearly not ridiculous. Measurements of the ratio of deuterium to hydrogen in Halley's comet are the same (within the measurement uncertainty) as in our ocean water. (We will return to this subject shortly.)

Mass Extinctions

Though scientists have concluded that the craters at Tunguska, in Arizona, and elsewhere were probably not produced by comets after all, they take very seriously the idea that comet bombardments have played an important role in the evolution of the earth. They may, indeed, have been responsible for mass extinctions.

Luis Alvarez and his colleagues have proposed that a large object—be it of asteroidal or cometary origin—hit the earth some 65 million years ago and caused the extinction of most of the species that existed at that time. Some such event seems to have occurred at the boundary between the Cretaceous and Tertiary periods (the K-T boundary). This boundary corresponds to the extinction of a variety of species, including the dinosaurs, that had flourished for the previous 140 million years. Though this boundary separates time periods, not geographical areas, it is clearly visible in the sediments deposited those millions of years ago. The clay layer between the Cretaceous and Tertiary sediments is very rich in iridium. Generally, iridium is a rare element in the earth's crust (one ten-thousandth of the amount in meteorites) because it shows an affinity for iron, and the primordial iridium was scrubbed out and carried down to the earth's core with the iron early in the formation of the earth.

A consistent explanation for the iridium-rich clay invokes a meteorite some 10 kilometers in diameter. Meteorites (and presumably all other bodies in the solar system) have a high iridium content, and the dust from an impact could be the source of the iridium in the clay. The impact would produce shock-heating of the atmosphere, changes in atmospheric chemistry, and acid rain. It would also produce an opaque dust blanket that would stay in the atmosphere for months. During that time, sunlight could not reach the ground or the ocean to produce photosynthesis. The food chain would be broken and mass starvation would be the result.

The idea is certainly plausible, but we tend to be distrustful of catastrophic explanations in physical science. The correct question really concerns the best explanation for the facts, whether catastrophic or otherwise. Is a single, large *bolide*—an especially bright meteor—required, or would a shower suffice? These ideas are currently being investigated and debated.

Other extinctions are evident in the geological record, and some are associated with iridium enhancements, such as the one near the Eocene-Oligocene boundary (35 million years ago). Could the meteorite scenario have occurred twice? Could it even be periodic? Extinctions during the last 250 million years have been described as occurring every 26 million years, and an intriguing cause has been proposed. A low-luminosity solar companion star—it has been dubbed Nemesis—could be in an eccentric orbit with a period of 26 million years. When Nemesis is near perihelion, it passes through the Oort cloud and the perturbation produces a comet shower. Over several million years a few would be expected to hit the earth.

This picture is consistent and exciting, and it is under discussion. There are some questions. Some scientists doubt the reality of the 26-million-year period. And do the bolides, if they are comets, have to reach the ground for the mechanism to work? Recall the Tunguska event and the fragility of the cometary nucleus. Perhaps, if the body is big enough, it (or major fragments) can reach the ground. Are there other events that could trigger periodic comet showers? A possible example is the sun's motion perpendicular to the plane of the Milky Way galaxy. The period of this motion is about 66 million years, so that the sun passes through the plane of our galaxy every 33 million years. Stars are denser near the galactic plane, and encounters with stars are more likely to produce comet showers from the Oort cloud.

The development of these ideas over the next few years should be very interesting. The interconnections between comet studies and other fields offer some of the best challenges to the comet scientist in this era.

Small Comets

Most of the comets that scientists study are roughly 1 kilometer in radius or larger. But is there a major population of comets that are so small that they cannot be seen by our current observational techniques? Probably not. We can imagine some lower limit to the size of active comets. If a comet is too small, the ices in the interior are not insulated enough to prevent rapid sublimation. However, there is probably a considerable range between the smallest possible comets and the size of those we normally observe. The hypothesis of a population of unobservable comets can be used to explain some phenomena in the solar system. This is an active, controversial, and exciting area of inquiry.

Satellite observations of the earth's upper atmosphere from above show localized changes in atmospheric brightness (Figure 8.9) that have been ascribed to a rather large influx of small comets. Cometesimals with radii in the range of a few meters to a few tens of meters have been postulated to explain some observations of hydrogen Lyman-alpha emission in the inner solar system. The hydrogen atoms originate by sublimation of the ices on these cometesimals. A cometesimal population large enough to produce the hydrogen Lyman-alpha emission is large enough to have produced all craters on the moon with diameters between 200 and 1500 meters over a period of 3.2 billion years on such areas as the Mare Tranquilli-

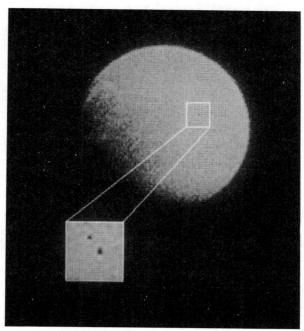

Figure 8.9 Atmospheric dark spots in an image of the earth's ultraviolet dayglow as observed from *Dynamics Explorer 1* on October 10, 1982. Some scientists hypothesize that these dark spots are produced by small comets that enter the earth's upper atmosphere. (Courtesy of L. A. Frank, University of Iowa)

tatis (Figure 8.10). These ideas are tantalizing, and exciting new results from research in these areas are expected.

The calculated numerical values of the size of the comet population necessary to explain these and other phenomena observed in the solar system vary widely and are not likely to settle down soon. The reason is that these values are inferred from a great variety of observations. Even if the population of small comets explains only a fraction of lunar craters or the Lyman-alpha background radiation, however, the number of

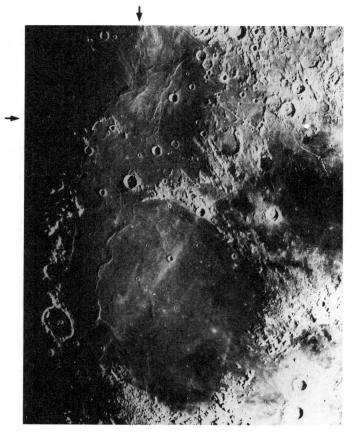

Figure 8.10 Mare Tranquillitatis, at top (arrows). Craters here can be used to estimate a cratering rate and, by implication, the number of comets that have hit the lunar surface. (Yerkes Observatory photograph)

small comets could still be very important in the physics of solar-system phenomena. Clearly a direct measurement of this population is necessary, and we can hope that it will be carried out in the near future.

Some evidence suggests that the atmospheres of the terrestrial planets—Venus, Earth, and Mars—might not have resulted from the original formation process. The ratio of deuterium to hydrogen (D/H) in water deep below the earth's surface, for example, is different from the value for ocean water, a situation that suggests that the ocean water may not have originated on the earth but have come from an outside source. Measurements of the deuterium/hydrogen ratio on Halley's comet by the *Giotto* investigators found a value essentially the same as that for ocean water. This striking similarity leads some scientists to believe that collisions with comets may have produced volatiles—atmospheres and oceans—on the terrestrial planets.

At present we have measurements for only one comet—Halley. If further research shows that the deuterium/hydrogen values on many other comets are close to the value for ocean water, the comet hypothesis may be taken seriously. Of course, this line of speculation is entirely compatible with the comet populations postulated to explain phenomena observed in the earth's upper atmosphere, the hydrogen Lyman-alpha observed in the inner solar system, and certain aspects of lunar cratering.

The abundance of deuterium implied by the deuterium/hydrogen ratio is a sensitive indicator of conditions for the formation of molecules or solids from their atomic constituents. Because the mass of the deuterium atom is twice that of the hydrogen atom, chemical reactions involving the two elements proceed at different rates. Empirically, the deuterium/hydrogen ratio varies widely with the astronomical circumstances. Note that we have not been talking about the creation or destruction of deuterium; it can be destroyed only in nuclear reactions. Rather, we have been considering the different rates at which hydrogen and deuterium form compounds. Thus continued, accurate measurements of D/H in comets, whether by in situ missions or by remote observations, are very important in cometary research.

Comets and the Origin of Life

If comets have supplied a significant fraction of the terrestrial atmospheres and oceans, they may also have supplied important quantities of the moderately complex organic molecules that could be the start of processes leading to the development of life on earth. Molecules such as methyl cyanide (CH_3CN) have been observed in comets, and formaldehyde (H_2CO) is an element in many models of the nucleus.

Measurement of polymerized formaldehyde [$(H_2CO)_n$] at Halley's comet in the mass range of 45 to 121 atomic mass units certainly supports comets as a potential source of organics that could speed up the evolutionary sequence outlined below. If the presence of polymerized formaldehyde is characteristic of all comets and if comets supplied a significant amount of organic material to the earth, comets may have contributed to the initiation of life on earth.

The general scenario for the origin of life follows the ideas introduced by Aleksander I. Oparin and J. B. S. Haldane in the 1920s. The earth's *primitive atmosphere*—its original atmosphere—is thought to have consisted of simple compounds of hydrogen, such as water, methane, and ammonia. The *hydrosphere* (the outer liquid portion of the earth, consisting primarily of the oceans) and the *lithosphere* (the earth's rocky shell, consisting of the crust and upper mantle) are also thought to have been made up of relatively simple compounds at the time the earth was formed. If these simple molecules were combined with an energy source (heat, ultraviolet light, or lightning, say), they would react to produce such substances as amino acids, proteins, and sugars. Subsequent laboratory experiments have verified that these reactions will occur. Collection of these simple organic substances in primordial shallow pools of water probably led to further chemical reactions that polymerized these simpler molecules into more complex organic substances—that is, substances that do not require biological processes for formation—of higher molecular weight. These substances, in turn, might have evolved to produce self-replicating molecules and ultimately the first living systems.

Comets may be an important source of terrestrial volatiles because of their specific elemental composition. Laboratory experiments have found that not all combinations of substances produce prebiotic molecules. To appreciate this concept, let us review the origin of the earth's atmosphere.

The earth's original atmosphere must have been lost quickly because the earth's gravity would be unable to keep the gases from the solar nebula (such as hydrogen and helium) from escaping. Thus the current atmosphere must be almost entirely of secondary origin. Possible sources of the secondary atmosphere from the interior of the earth are volcanoes, which erupted at an essentially constant rate over geological time, and radioactive decay. Possible sources from the exterior are comets—there must have been many more of them in the past, before most of them were swept up by the planets—and meteoroids. The composition of the atmosphere can also be modified by chemical reactions with the land surface and the oceans (the oceans contain vast amounts of carbon dioxide in solution) and by photosynthesis. Here, of course, we are interested in the atmosphere that existed when the prebiotic molecules were in the process of formation.

Table 8.1 shows the percentages of carbon, hydrogen, oxygen, and nitrogen in various atmospheric sources and some biological entities. CI chondrites—a type of stony meteorite with a high relative abundance of carbon—were chosen to represent meteroids because of all meteorite classes they contain the largest fraction of volatiles. The close match between the abundances of CHON molecules in comets and in living ma-

Table 8.1 Abundances of Carbon (C), Hydrogen (H), Oxygen (O), and Nitrogen (N) in Various Sources and Objects (percent of atoms)

Source or object	C	H	O	N
Volcanic gases (Kilauea and Mauna Loa)	5	55	37	3
Meteorites (CI chondrites)	7	15	74	5
Comets	10	56	31	3
Bacteria	6	63	29	1
Mammals	10	61	26	2

Data adapted from A. H. Delsemme, in C. Ponnamperuma, ed., *Comets and the Origin of Life* (Dordrecht, Holland: Reidel), pp. 141–159, with additions adapted from J. Gilluly, A. C. Waters, and A. O. Woodward, *Principles of Geology*, 3rd ed. (New York: W. H. Freeman), p. 579.

terial is clearly evident. The volcanic gases are reasonably close to the life forms, but they appear to be deficient in carbon. The vital contribution of comets to the mixture could be to supply the necessary carbon.

The fossil evidence has pushed the origin of terrestrial life back to at least 3.5×10^9 years ago and probably to 4.0×10^9 years ago. During the early stages of the solar system, collisions with debris left over from the formation process, including comets, must have occurred at a rate much higher than they do now. Comets carrying already complex hydrocarbons would certainly enhance the evolutionary process and perhaps were the mechanism that set it in motion. If the comets condensed from dense interstellar material, the existence of complex molecules in them implies that the reactions that produced these molecules are widespread in the cosmos. The recent report of the C_{60} molecule in interstellar space (along with the long list of molecules already known) lends strong support to this view.

The Future

The tale we have told so far is a good news–good news story. The first good news is that the well-coordinated series of research activities devoted to comet Halley has revolutionized our understanding of comet Halley itself and of comets in general. The second good news is for all of us who enjoy the search for knowledge: the studies of comet Halley simply scratched the surface. It may seem inconsistent to say that studies that revolutionized our understanding simply scratched the surface, but it is true. There are still many fascinating unanswered questions to be pursued.

Recent Comets

There *is* life in comet science after the 1985–1986 apparation of Halley's comet. The appearance of each additional comet is a spur to further research.

Figure 9.1 shows a wide-field view of comet Bradfield on October 20, 1987. The contortions of the tail well away from the head are indicative of large and interesting changes in the solar-wind speeds. Figure 9.2 shows a wide-field image of comet Brorsen-Metcalf on September 7, 1989. This comet somewhat resembles comet Halley because it has a period of 72 years. In contrast to Halley, though, Brorsen-Metcalf has a direct (prograde) orbit.

The flight of a sounding rocket on April 28, 1990, produced a spectrum of comet Austin from 910 to 1180 angstroms; the exposure time was 258 seconds and the resolution was approximately 3 angstroms (Figure 9.3). The comet was about 0.6 astronomical unit from both the sun and the earth. The principal emissions are from hydrogen and oxygen. Scientists expected the hydrogen emission at 1026 angstroms (Lyman-beta), from excitation of the comet's hydrogen cloud by solar Lyman-beta emission. The oxygen lines are produced by a chance coincidence between solar Lyman-beta emission and an energy level in the oxygen atom. Different transitions from this level produce emissions at 1026, 1042, and 1128 angstroms.

At least as interesting were the emissions not seen. Argon has transitions at 1048 and 1066 angstroms, and the helium transition at 584 angstroms should appear in the second order at 1168 angstroms. Comet Austin's orbit indicates that it was making its first pass through the inner solar system. Thus the absence of argon and helium emissions in this comet may place

Figure 9.1 Comet Bradfield (1987s) on October 20, 1987. The distance from the head to the large kink is approximately 4 degrees. (JOCR photograph by D. A. Klinglesmith III)

Figure 9.2 Comet Brorsen-Metcalf (1989o) on September 7, 1989. (Courtesy of Chris Schur, Payson, Arizona)

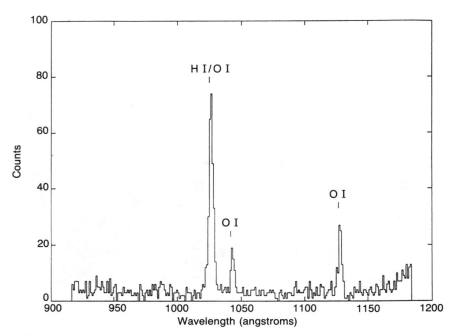

Figure 9.3 Spectrum of comet Austin (1988c₁) from 910 to 1180 angstroms. (Courtesy of James C. Green, Center for Astrophysics and Space Astronomy, University of Colorado, Boulder)

important constraints on theories explaining the formation and evolution of comets.

Finally, we have begun to study comets with the Hubble Space Telescope. Figure 9.4 shows a 4-second exposure of comet Levy obtained with the Wide Field and Planetary Camera on September 27, 1990. The infrared image shows the inner coma with a dust fan in the southward direction. The resolution of the processed image is approximately 75 kilometers. Sophisticated image enhancement may reveal additional features of comet Levy.

Observations of these comets may yield new insights into comets. We stress that results from comet Halley dominate our current view, but we think it's unwise to assume that Halley is representative of all comets.

Outburst on Comet Halley

The extraordinary effort to observe Halley's comet at all phases of its recent apparition continues to produce interesting results. On the morning of February 12, 1991, the comet was about 300 times brighter than expected and appeared as an

Figure 9.4 Image of comet Levy (1990c) on September 27, 1990, obtained with the Wide Field and Planetary Camera on the Hubble Space Telescope. This 4-second exposure in the infrared shows a fan-shaped region on the sunward side of the nucleus; this feature consists of dust carried along by gas flowing away from the nucleus. (Space Telescope Science Institute)

extended nebula (Figure 9.5). Observations showed that the nebular material is composed entirely of dust grains. There was no obvious change in the appearance of the nebula over a five-day period. The puzzling aspect of this outburst is that it occurred when the comet was 14.3 astronomical units from the sun—that is, about halfway between the orbits of Saturn and Uranus. What energy source could produce such an outburst so far from the sun?

One explanation is that energy might have been released from the comet's interior due to phase changes in ice. These phase changes are unlikely in Halley's comet, however, because the comet has been heated repeatedly; moreover, the surface temperature of the nucleus would be roughly 70 K, a

Figure 9.5 Outburst of February 12, 1991, on Halley's comet, as observed by O. Hainaut and A. Smette. The comet was over 2 billion kilometers from the sun. The central part of the dust cloud is over 300,000 kilometers across. (European Southern Observatory)

value far below the temperatures associated with the ice phase transitions.

Chemical reactions are another possible explanation. Extremely volatile gases could be stored in the comet's interior and could have come into contact with each other as a result of some other event at the comet, producing a chemical reaction with a release of energy. This mechanism seems very unlikely, though, because the nebula was observed to be made up entirely of dust.

A sudden change in solar-wind conditions at the comet could disrupt the charge balance and release a cloud of dust (see Chapter 6). This mechanism is quite plausible. A solar flare could produce a shock wave in the solar wind. This might crack the surface and allow dust to escape.

Last, the comet may have collided with a single object or a swarm of objects in the outer solar system. The likelihood of this happening depends on the population of small bodies between Saturn and Uranus. Also, such collisions would have had to produce a steady-state condition lasting over a period of days.

This new outburst phenomenon may lead to vital new understanding of the cometary nucleus. Observations will be intensified, and the theories will be reexamined in light of new data.

Very Large Comets?

Do we have evidence that very large comets exist—comets much larger than the ones we typically study, which seldom measure more than 10 kilometers? Current theories of comet formation contain nothing that would preclude the formation of large comets. Generally, we would expect that the larger the comets, the fewer of comparable size there would be; that is, there would be many more small comets than large ones.

A very large comet possibly was the progenitor of the Kreutz family of sun-grazing comets. These comets, as we saw in Chapter 2, undoubtedly result from the disruption of a large comet, possibly the very bright comet that Aristotle saw in 371 B.C.

Recent studies of the object 2060 Chiron may also indicate the existence of large comets. Chiron, the most distant known asteroid, has an orbit ranging from 8.5 to 18.9 astronomical units; that is, between the orbits of Saturn and Uranus. If the albedo of an object is estimated from color measurements, the brightness can be used to determine the object's size. The radius of Chiron has not been determined very accurately, but it is known to be in the range of 65 to 200 kilometers. Currently Chiron is moving toward perihelion passage in 1996.

Sometime between late 1986 and early 1988, Chiron doubled in brightness. Additional observations indicated that the brightening was sporadic. As we saw in Chapter 3, the cause of cometary outbursts is not understood.

Analysis of Chiron's orbit indicates that it is unstable on time scales of several thousand years. In addition, this instability is chaotic, meaning that very small changes in the current position produce very large changes in the future position. Simply put: (1) Chiron will be thrown out of its orbit sometime during the next few thousand years; (2) we cannot predict when the change will happen or where Chiron will go; and (3) Chiron cannot have been in its current orbit very long.

Various mechanisms have been proposed to explain the brightening, but the most plausible is outgassing as a result of sublimation of volatiles caused by heating. At Chiron's distances from the sun, water (H_2O) does not sublimate, but other volatiles such as methane (CH_4), carbon monoxide (CO), carbon dioxide (CO_2), and molecular nitrogen (N_2) do. If Chiron had been in its present orbit since the birth of the solar system, these volatiles would have been lost long ago.

Chiron's orbital situation and apparent intermittent release of volatiles support the view that Chiron has been deposited into its present orbit recently, and before that event was stored beyond 20 astronomical units from the sun. Thus Chiron may be a very large comet from the Kuiper-Duncan disk or the inner Oort cloud.

Further observations are needed to establish Chiron's properties and to see if other Chiron-like objects can be found. However, the apparent discovery of large (roughly 100-kilometer) objects recently sent inward from the Oort cloud opens new opportunities for probing the circumstances of the formation of the solar system and the origin of comets.

Missions to Comets

The history of space missions to comets, particularly with respect to the U.S. program, has been frustrating. Missions to comets can be placed in three broad classes. During a *flyby*, a spacecraft flies past a comet, usually at high speed, and makes close observations; during a *rendezvous mission*, a spacecraft flies into the vicinity of a comet and may orbit alongside it for a time; and during a *sample return*, a spacecraft lands on the nucleus of a comet, scoops up a sample of material, and returns with it to earth.

NASA mission studies of comet Halley began in the 1970s, and until late 1979 it was thought to be possible to rendezvous with Halley's comet at its 1985–1986 appearance. The plan required development of an advanced solar-electric propulsion system called an "ion drive," but approval for its development on a rush basis was not received. Although alternative mission plans were developed, no *new start* (as NASA calls a congressional go-ahead to start a new space mission) was received for a NASA mission to Halley's comet. The comet physicist Paul Feldman summarized the situation succinctly: "Planning for the now nonexistent NASA mission to Halley was begun as early as 1970."[1]

The alternative mission plans initially focused on a mission to comets Halley and Tempel 2. The flight dynamicists had discovered that an ion-drive mission could fly by Halley's comet and then rendezvous with comet Tempel 2. This mission evolved into the International Comet Mission (ICM) through collaboration with the European Space Agency (ESA), which was to supply a spacecraft that would be dropped off the main spacecraft to probe Halley's coma. The Halley encounter would have occurred in November 1985, and the Tempel 2 rendezvous would have been from July 1988 to July 1989. This marvelous mission would have explored two comets and launched the space exploration of comets with a flourish. The United States's part of the mission expired early in the game.

---

1. Paul Feldman, *Science* 219, no. 4583 (1983):347–354.

NASA's last attempt was the Halley Intercept Mission, a simpler, less expensive enterprise. Ion drive was not needed and excellent science would still have resulted. A feature of this mission was a superb imaging system that would have produced high-quality images of the nucleus with a narrow-angle system and an almost continuous record of the comet's large-scale structure through the use of a wide-angle system during an extended observation phase. The demise of this mission during the summer of 1981 not only was a blow to comet researchers but produced something of a public outcry as well.

A last-ditch effort was the Halley Fund, organized by a group of space enthusiasts who hoped to fund a mission through contributions. This try, unfortunately, failed like all other U.S. attempts.

Fortunately for comet science, the exploration of Halley's comet by the flyby missions of the Halley Armada was carried out by the European Space Agency, the Soviet Union, and Japan. The U.S contribution was the retargeted *ICE* spacecraft sent through the tail of comet Giacobini-Zinner and then, as a distant probe, to comet Halley. (As a historical note, the *Giotto* spacecraft had its origins as part of the Joint International Comet Mission.) In addition, there are opportunities to retarget some of the spacecraft in the Halley Armada to other comets.

Space missions to comets are extremely valuable, and many facts about comets can be established in no other way. Yet these missions, whether flyby, rendezvous, or sample return, have limitations, too. The flyby usually supplies a snapshot of a single comet. The rendezvous supplies measurements over an extended period of time and over different phases of activity, but for only one comet. The sample return, too, applies to a single comet. Because each comet is in some ways unique, generalizing from a small sample may be misleading. This does not mean that direct exploration is futile, merely that we should look for simple and inexpensive (by space-science standards) ways to expand the sample of space-probed comets.

Three potential missions address this situation. In the United States we have the Multi-Comet Mission (MCM); in Europe ESA's Comet Atmosphere Encounter and Sample Return (CAESAR) mission, which has many similarities to the Multi-Comet Mission, is in the conceptual stage; and in Japan a mission called SOCCER is being studied.

The CAESAR mission features a slow flyby (less than 10 kilometers per second) of a short-period comet with scientific instruments similar to those aboard *Giotto,* plus the ability to collect samples in situ. The *Giotto*-like instrumentation would consist of an imaging system, optical and mass spectrometers, dust detectors, and plasma instruments, including a magnetometer. Results from these instruments should allow a detailed comparison of a short-period comet and a highly active one (Halley).

The in situ collection and sample-return part of the mission involves techniques for collecting volatile and nonvolatile components. Volatiles can be collected on chemically inert surfaces (such as aluminum foil) and chemically active surfaces (such as coated foils); the chemically active surfaces are necessary for collection of hydrogen, carbon, nitrogen, oxygen, and molecules made up of these elements. The collection of the nonvolatiles—that is, dust—focuses on decelerating the dust particles through a tenuous medium (such as foam). Laboratory tests indicate that a significant fraction of the dust particles can be recovered intact or as large fragments. In either case, the mineralogy can be preserved.

The mission profile involves launch of the combined transfer spacecraft and return module to the target comet, deployment of the collection devices at the comet, retraction of the collection devices into the return module, return of the combined spacecraft to the vicinity of earth, separation of the return module from the transfer spacecraft, direct atmospheric reentry of the return module, and deployment of a parachute system for the final descent. CAESAR's approach distance to the cometary nucleus is expected to be about 50 kilometers. There are many candidate comets for this mission.

The Multi-Comet Mission study envisions encounters in the velocity range of 10 to 12 kilometers per second, and in fact is planned as a tour of comets and asteroids during the years 1994 to 2005. Two spacecraft are to be used. The *Observer* spacecraft can accomplish the tour with earth-swingby maneuvers and propulsion systems. The scientific instruments for cometary observations include an imaging system; a mass spectrometer to measure the masses of molecules, neutral atoms, and ions; a dust analyzer; and instruments for plasma physics experiments, including a magnetometer to measure local magnetic fields. Additional instruments would permit solar studies and studies of interplanetary physics. The comet instruments and an X-ray spectrometer can be used to study the asteroids encountered in this mission.

The separate *Probe* spacecraft (Figure 9.6) would deploy dust collectors during encounters with comets. Its propulsion system and an aerobrake assembly would permit a return to low earth orbit. The aerobrake is a device being developed by NASA to use the resistance of the residual atmosphere at orbital altitudes to alter the motion of a spacecraft.

The Multi-Comet Mission is summarized in Figure 9.7. Both spacecraft would be launched into an earth-return orbit. *Observer* would arrive at the target a few days before *Probe* to provide navigational assistance for the collection phase. After being launched in January 1994, *Observer* would reach comet Tempel 1 in June 1994, the asteroid Hestia in May 1998, comet Tempel 2 in August 1999, comet Encke in November 2003, and the asteroid Eros in November 2005. After each encounter, one or two earth-swingby maneuvers would be performed.

The *Probe* spacecraft would return to earth after the encounter with comet Tempel 1 and would be placed in low earth orbit. The dust sample would be recovered, and *Probe* would be refurbished and sent to comet Tempel 2 and beyond. *Probe* would not be sent to comet Encke.

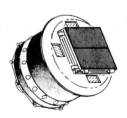  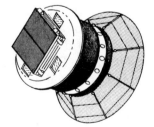

Probe is launched and cruises to comet

Probe deploys dust collector panels just before cometary encounter

Probe closes and seals dust collector panels after encounter

Aerobrake is deployed after return to earth orbit

Figure 9.6 The operation of the Multi-Comet Mission *Probe* spacecraft. (Courtesy of R. W. Farquhar, NASA-Goddard Space Flight Center)

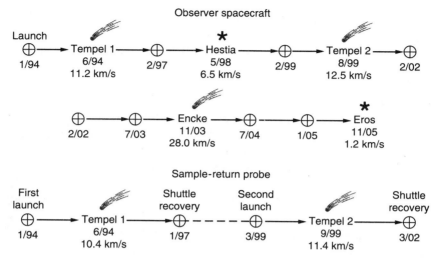

Figure 9.7 Summary of the Multi-Comet Mission. (Courtesy of R. W. Farquhar, NASA-Goddard Space Flight Center)

The potential return from this mission is impressive. Extensive data from three more comets would give a tremendous boost to comparative studies of comets. The Multi-Comet Mission seems to have an edge over CAESAR in overall science return, but CAESAR has the edge over the MCM in having a lower flyby speed, which significantly improves the probability that intact dust would be collected.

Currently the MCM and CAESAR missions are not being actively pursued. The concepts involved in these missions, however, are being studied jointly by the Japanese Space Agency (ISAS) and NASA. The Joint NASA/ISAS mission is called SOCCER, for Sample of Comet Coma Earth Return. If this mission is realized, the spacecraft will be launched in 1995, fly by comet Churyumov-Gerasimenko in January 1996, and return to earth in the year 2000.

The rendezvous phase of the cometary exploration sequence can be carried out by NASA's Comet Rendezvous Asteroid Flyby (CRAF) mission. The approach here is simply to rendezvous with a comet and fly along with it for an extended period of time while the comet passes through its phases of activity. The execution is not so simple. The mission is long and the scientific objectives require a large, fairly complex spacecraft. The science includes essentially everything appropriate, from detailed imaging of the nucleus to chemical compositions and plasma physics. The long time that the spacecraft spends near the comet, including the time when activity is low, should allow the comet's mass to be determined from celestial mechanics calculations. A novel feature of the CRAF mission is the *penetrator,* a projectile with scientific instruments to be fired into the comet's surface rather like a harpoon. The penetrator will provide information on the bulk properties of the subsurface layers and data on the composition, temperature, and flow of energy. Data will be telemetered back to the main spacecraft for about ten days. The *CRAF* spacecraft (Figure 9.8) near a cometary nucleus is shown in an artist's rendering in Plate 14.

The specific targets currently are comet Kopff and the asteroid Hamburga. Launch is expected in August 1995. The trajectory to comet Kopff is shown in Figure 9.9. The spacecraft makes a flyby of the asteroid Hamburga in January 1998 and arrives at Kopff in August 2000, some 850 days before perihelion. Penetrator delivery is in July 2001. The length of time to nominal end of mission in 2003 is 964 days; the mission may be extended until fuel runs out.

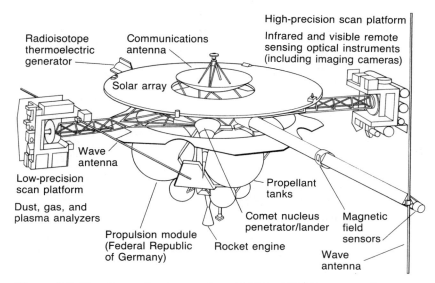

Figure 9.8 The *Comet Rendezvous Asteroid Flyby (CRAF)* spacecraft. (NASA-Jet Propulsion Laboratory)

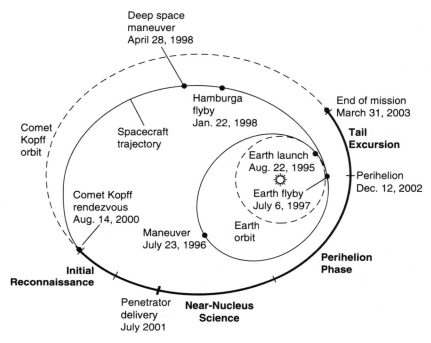

Figure 9.9 The *CRAF* orbit to comet Kopff and the asteroid Hamburga. (NASA-Jet Propulsion Laboratory)

The CRAF mission offers tremendous science return and constitutes a vital phase in the orderly exploration of comets. The problems to be overcome involve politics and resources.[2]

Among the comet missions currently being planned, ESA's Rosetta mission is the most extensive. This mission contemplates a landing on the surface of a comet to obtain a sample of the nuclear materials and return it to earth for sophisticated laboratory analysis. Many aspects of the mission are still under study, but it may resemble the scenario shown in Figure 9.10. The spacecraft may use solar-electric propulsion. Special study is needed of the navigational requirements for landing on the comet's surface, for obtaining the sample, and for reentering the earth's atmosphere.

The Rosetta mission would carry out the sample-return phase of cometary exploration. Because it is a "Cornerstone" mission in ESA's long-term Horizon 2000 program, its ultimate completion is likely.

---

2. As of February 1992, CRAF is officially cancelled.

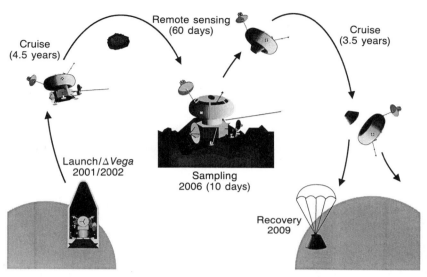

Cruise
(4.5 years)

Remote sensing
(60 days)

Cruise
(3.5 years)

Launch/Δ Vega
2001/2002

Sampling
2006 (10 days)

Recovery
2009

Figure 9.10   Overview of the Rosetta mission for Comet Nucleus Sample
Return. (Courtesy of G. Schwehm, Space Science Department,
European Space Agency)

Comet Data

One of the major achievements of the International Halley
Watch has been the production of an extensive data archive
documenting the latest apparition of Halley's comet. The archive includes a major fraction of the data obtained both from
the space missions and from the ground. Clearly this archive
will be a source of research material for investigators for years
to come.

A pressing need as the exploration of comets continues is
to archive all new data on cometary research. The creation of
a successor organization to the International Halley Watch, to
be called the International Comet Watch, has been proposed.
This suggestion has considerable merit; such an organization
would foster the building and maintenance of a continuing data
base on comets, which would be made available to all researchers in the field. The suggestion, in essence, has been incorpo-
rated into NASA's Planetary Data System, which is beginning
operation.

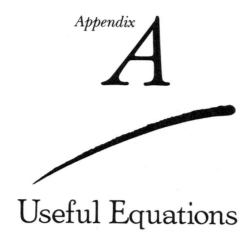

Appendix
A

Useful Equations

Universal Gravitation and the Orbits of Comets

The calculation of orbits rests on Newton's second law, which states that

$$F = ma \qquad (1)$$

where F is the force, m is the mass of the body, and a is the acceleration. The force for bodies orbiting the sun is given by the law of universal gravitation, or

$$F = \frac{GM_\odot m}{r^2} \qquad (2)$$

Here G is the gravitational constant (6.7×10^{-8} dyne cm^2 gram^{-2}), $M_\odot$ is the mass of the sun (2.0×10^{33} grams), and r is the distance from the sun to the body. The solutions to these equations are the basis for calculating orbits in the solar system.

The basic concepts of orbits are easily expressed with the aid of elementary mathematics. When there are no planetary perturbations or nongravitational forces, equations (1) and (2) are appropriate and a small mass describes an orbit around the sun that is a conic section with the sun at one focus. The essential results include the equation for conservation of energy and Kepler's laws.

Conservation of energy states that the total energy E, the sum of the kinetic and gravitational potential energies, is constant, or

$$E = \frac{v^2}{2} - \frac{GM_\odot}{r} \qquad (3)$$

The kinetic energy per unit mass is given by $v^2/2$, where v is the velocity (in centimeters per second). The gravitational potential energy per unit mass is given by $-GM_\odot/r$.

The orbit depends on the value of the total energy E. If E is negative, the body is bound to the sun and the orbit is an ellipse. If E is positive, the body is not bound to the sun and the orbit is a hyperbola. If E is zero, the body falls toward the sun, starting with zero velocity at infinity, and the orbit is a parabola.

These differences can also be seen by considering the general equation for a conic section in polar coordinates,

$$r = \frac{q(1+e)}{1+e \cos \theta} \qquad (4)$$

Here r is the distance from the focus containing the sun, q is the perihelion distance, θ is an angle measured from perihelion, and e is the eccentricity. Specifying the eccentricity e determines the particular conic section and the total energy E (see Table A.1).

The different orbits are illustrated in Figure 2.9, where the ellipse has $e = 0.9$, the parabola has $e = 1.0$, and the hyperbola has $e = 1.1$.

For an elliptical orbit, the perihelion distance q is given by

$$q = a(1-e) \qquad (5)$$

Table A.1 Energy and Eccentricity of Orbit Shapes

Shape	Energy	Eccentricity
Ellipse	$E < 0$	$e < 1$
Parabola	$E = 0$	$e = 1$
Hyperbola	$E > 0$	$e > 1$

and the aphelion distance Q is given by

$$Q = a(1 + e) \tag{6}$$

For an ellipse, equation (4) can be written as

$$r = \frac{a(1 - e^2)}{1 + e \cos \theta} \tag{7}$$

Kepler's laws can be stated as follows: The semimajor axes a are related to the periods P by Kepler's third law:

$$a^3 = P^2 \tag{8}$$

where a is in astronomical units and P is in years.

The law of areas is given by

$$\frac{dA}{dt} = \frac{1}{2} r^2 \frac{d\theta}{dt} = \frac{h}{2} \tag{9}$$

where r and θ are the polar coordinates as above, dA/dt is the rate of change of area A over time, and h is a constant.

The speed of a comet in a parabolic orbit is given by setting $E = 0$ in equation (3), and we find

$$v = \left(\frac{2GM_\odot}{r}\right)^{1/2} \tag{10}$$

For $r = 1$ AU, this speed is 42 kilometers per second. Thus equation (10) can be rewritten as

$$v = \frac{42 \text{ km/s}}{r^{1/2}} \tag{11}$$

where r is in astronomical units.

Nongravitational Forces

The treatment of nongravitational forces adds the appropriate terms to the equation of motion in parametric form. A fit to observed motions then determines the nongravitational parameters. The equation of motion becomes

$$\frac{d^2\mathbf{r}}{dt^2} = \frac{-\mu\mathbf{r}}{r^3} + \frac{\partial R}{\partial \mathbf{r}} + A_1 g(r)\hat{r} + A_2 g(r)\hat{T} \qquad (12)$$

Here $\mathbf{r}$ is the radius vector; $\mu = GM_\odot$; R is the planetary disturbing function; $\hat{r}$ and $\hat{T}$ are the radial and tangential unit vectors, respectively; A_1 and A_2 are the radial and tangential nongravitational parameters, respectively; and $g(r)$ is a function proportional to the rate of sublimation of ices. Traditionally,

$$g(r) = \alpha \left(\frac{r}{r_0}\right)^{-m} \left\{ 1 + \left(\frac{r}{r_0}\right)^n \right\}^{-k} \qquad (13)$$

where α is a normalization parameter and m, n, k, and r_0 are parameters determined by the particular ice; r_0 is the distance beyond which sublimation falls off rapidly (it is 2.8 AU for water ice).

Until recently, everything in these equations was known except A_1 and A_2; these were determined by fitting an orbit to the observations. Note, however, that equation (13) is symmetrical with respect to perihelion. This has the effect of reducing the importance of the radial term A_1. Donald Yeomans and Paul Chodas (1989) have introduced forms for $g(r)$ that can be asymmetric with respect to perihelion. Their parameter AT gives the number of days before or after perihelion at which the maximum water-ice sublimation occurs. Thus, in evaluating $g(r)$, they use $g(r')$, where r' is the value of r at $t' = t - AT$. This approach yields results in which the radial term A_1 is the most significant.

Perfect Gases, Thermal Speed, and Sound Speed

The relationship between the pressure P, volume V, number N of atoms or molecules, and temperature T is called the equation of state. The approximation called a "perfect gas" is valid for cometary applications. We have

$$PV = NkT \qquad (14)$$

Here P is the pressure in dynes per square centimeter, V is the volume in cubic centimeters, N is the number of particles, k is Boltzmann's constant (1.4×10^{-16} erg/K), and T is the temperature in degrees Kelvin. This equation is often written as

$$P = nkT \qquad (15)$$

where $n = N/V$ is the number of particles per cubic centimeter.

The root-mean-square or thermal speed in a gas is given by

$$v_t = \left(\frac{3kT}{m}\right)^{1/2} \qquad (16)$$

where m is the average mass of the particles.

The speed of sound is given by

$$v_s = \left(\gamma \frac{kT}{m}\right)^{1/2} \qquad (17)$$

where γ is the ratio of the specific heat at constant pressure to the specific heat at constant volume. For monoatomic, ideal gases, $\gamma = 5/3$ and

$$v_s = \left(\frac{5}{3} \frac{kT}{m}\right)^{1/2} \qquad (18)$$

The value of γ decreases for diatomic gases to approximately 1.4. Disturbances in a nonionized gas travel at the speed of sound; compare with the Alfvén speed (below).

Simple Plasma Physics

The conditions necessary for plasma behavior (discussed in Chapter 4) are as follows:

1. The Debye length is small in comparison with the characteristic dimension of the system. The Debye length in centimeters is given by

$$\lambda_0 = 4.9 T^{1/2} N^{-1/2} \qquad (19)$$

For conditions appropriate to the tail of comet Giacobini-Zinner, we have $T = 10^4$ K and $N = 50$ cm^{-3}. Here $\lambda_0 = 70$ cm, a value much smaller than any characteristic dimension for cometary plasma.

2. A sphere with radius of 1 Debye length contains many electrons. This condition is clearly met for $\lambda_0 = 70$ cm and $N = 50$ cm^{-3}.

3. The plasma is approximately neutral, meaning no net charge per unit volume. See the discussion in Chapter 4 and the discussion for condition 4.

4. Plasma oscillations are not strongly damped by collisions. Plasma electrons can oscillate collectively around the massive ions. They do so at the plasma frequency (in hertz, Hz)

$$\nu_p = 9 \times 10^3 N^{1/2} \qquad (20)$$

For $N = 50$ cm^{-3}, ν_p is 60 kHz. The classical electron-ion collision frequency in hertz is given by

$$\nu_c = 50 N T^{-3/2} \qquad (21)$$

For $N = 50$ cm^{-3} and $T = 10^4$ K, $\nu_c = 2.5 \times 10^{-3}$ Hz. Thus collisions occur infrequently compared to the plasma frequency, and this condition for plasma behavior is satisfied.

Charged particles spiral around a magnetic field with the Larmor radius

$$r_L = \frac{m v_\perp c}{ZeB} \qquad (22)$$

where $v_\perp$ is the velocity perpendicular to the magnetic field B, m is the mass of the particle, c is the speed of light (3.0×10^{10} cm/s), Z is the charge ($= 1$ for singly charged particles), $e = 4.8 \times 10^{-10}$ electrostatic unit (esu), and B is in gauss. For thermal speeds of $T = 10^4$ K and $B = 2 \times 10^{-4}$ gauss, the Larmor radii for electrons and ions ($m = 20$ AMU) moving at an angle of 45° to the field lines are about 0.1 kilometer and 25 kilometers, respectively.

The Alfvén speed,

$$V_A = \frac{B}{(4\pi\rho)^{1/2}} \qquad (23)$$

determines the speed at which disturbances propagate in a magnetized plasma. The density ρ is given by the average mass in atomic mass units times the mass of the hydrogen atom times the number density. For the parameters mentioned here, $V_A = 12$ kilometers per second.

Light and the Doppler Effect

Light can be described as exhibiting the properties of a wave or of a particle. In either case, light in a vacuum (and in most astronomical circumstances) travels at speed c. The wave has wavelength λ and frequency ν given by

$$\lambda\nu = c \qquad (24)$$

Because c is a constant, equation (24) shows that as λ increases, ν decreases, and vice versa.

When light is considered as a particle, the energy of the individual photons is given by

$$E = h\nu \qquad (25)$$

where h is Planck's constant (6.6×10^{-27} erg s). Thus, in view of equations (24) and (25), a higher frequency ν means a higher energy E but a lower wavelength λ.

The Doppler effect for an object moving directly toward or directly away from the observer with speed v is given by

$$\frac{\Delta\lambda}{\lambda} = \frac{v}{c} \qquad (26)$$

Here c is the speed of light, λ is the rest wavelength of a spectral line, and $\Delta\lambda$ is the shift of the spectral line due to the motion v. The shift is positive (to longer wavelengths) when the motion is away from the observer and negative (to shorter wavelengths) when the motion is toward the observer.

Inverse-Square Law of Brightness

The inverse-square law of brightness is written very simply as

$$I = I_0 \left(\frac{r_0}{r}\right)^2 \tag{27}$$

where I is the brightness or intensity at arbitrary distance r and I_0 is the brightness at the reference distance r_0.

Astronomical Magnitudes

Astronomers use a magnitude scale derived from the naked-eye description of the brightest stars as first magnitude and the faintest stars visible as sixth magnitude. For stars with brightnesses b_1 and b_2, their magnitudes H_1 and H_2 are related by

$$H_2 - H_1 = 2.5 \log_{10} \frac{b_1}{b_2} \tag{28}$$

Note that objects that differ by a factor of 100 in brightness differ by 5 magnitudes.

The Law of Cometary Brightness

The law of cometary brightness can be written as

$$J = J_0 f(\Delta) F(r) \tag{29}$$

where J_0 is a reference brightness and the functions $f(\Delta)$ and $F(r)$ represent the variations due to the distance between the earth and the comet, Δ, and the distance between the sun and the comet, r. The function $f(\Delta)$ is simply the inverse-square law just described, and so

$$J = \frac{J_0}{\Delta^2} F(r) \tag{30}$$

The heliocentric variation could contain many factors and was represented by a function proportional to r^{-n}, where the parameter n was determined empirically. Then we have

$$J = \frac{J_0}{\Delta^2 r^n} \qquad (31)$$

Cometary brightnesses are often expressed in terms of magnitudes, and equation (31) becomes

$$H = H_0 + 5 \log \Delta + 2.5n \log r \qquad (32)$$

If we also remove the variation due to distance from earth by defining

$$H_\Delta = H - 5 \log \Delta \qquad (33)$$

we have

$$H_\Delta = H_0 + 2.5n \log r \qquad (34)$$

Thus, according to equation (34), a plot of cometary magnitude versus $\log r$ should be a straight line with slope $2.5n$. The average value for n is approximately 4. If the comet did not change size with varying distance from the sun, the amount of reflecting or scattering material would remain the same, and the brightness would follow the inverse-square law, or $n = 2$. The average value of $n = 4$ means that the amount of cometary material increases as the comet approaches the sun. In practice, values of n from 2 to 6 are common, and the extremes are -1 to 11.

These magnitudes refer to the total magnitude of the comet. Magnitudes referring only to the central concentration

Table A.2 Visual Photometric Parameters of Eight Comets

Comet	H_0	n
Bradfield 1974 III	7.61	2.92
P/Forbes[a]	10.40	4.00
P/Honda-Mrkos-Pajdušáková 1974 XVI	10.62	2.93
Kobayashi-Berger-Milon 1975 IX	7.34	3.77
Bradfield 1975 XI	8.88	2.91
West 1976 VI	5.94	2.42
Meier 1978 XXI		
$\quad 3.00 > r > 2.14$	-0.22	6.64
$\quad 1.20 < r < 4.23$	2.70	3.85
P/Stephen-Oterma 1980 X	3.46	11.92

[a]The value $n = 4$ was assumed, and the value H_0 was calculated.

of light are called nuclear magnitudes, although we emphasize that they do not represent the magnitude of the nucleus itself. The values of visual photometric parameters of a few comets are given in Table A.2.

Radiation Laws

The properties of an ideal radiator or black body are useful in many astronomical applications. A black body radiates at an intensity per frequency interval given by

$$B_\nu(T) = \frac{2h\nu^3}{c^2} \frac{1}{e^{h\nu/kT} - 1} \qquad (35)$$

where h is the Planck constant, c is the speed of light, k is the Boltzmann constant, ν is the frequency, and T is the temperature.

The total amount of energy emitted by a square centimeter of a black body is given by

$$E = \sigma T^4 \qquad (36)$$

where σ is the Stefan-Boltzmann constant (5.7×10^{-5} erg cm^{-2} deg^{-4} s^{-1}).

Finally, the maximum radiant energy is emitted at a wavelength in centimeters given by Wien's law, or

$$\lambda_{max} = \frac{0.29}{T} \qquad (37)$$

Equation (37) can be used to estimate the wavelength region that is important for radiation at various temperatures.

Energy Balance for Comets

The surface of a comet receives energy from sunlight, which goes into heating the material, sublimating ices, or being conducted into the interior. In this simple discussion, we neglect the effects of conduction into the interior. The basic equation is

$$F_0(1-A_0)\frac{\cos\theta}{r^2}=(1-A_1)\sigma T^4+Z(T)L(T) \qquad (38)$$

The various terms represent

$$\left\{\begin{matrix}\text{Energy received}\\\text{from the sun}\end{matrix}\right\}=\left\{\begin{matrix}\text{Radiation back}\\\text{to space}\end{matrix}\right\}+\left\{\begin{matrix}\text{Vaporization}\\\text{of ices}\end{matrix}\right\}$$

Note that the radiation back to space is $(1-A_1)\,\sigma T^4$ for a gray body; for a black body $A_1=0$, and the expression is the same as equation (36). If everything else is fixed, the temperature T changes to balance this equation. Here F_0 is the solar constant or solar flux at earth (2.0 cal cm^{-2} min^{-1}); A_0 is the albedo in visual wavelengths (approximately 5000 angstroms) where the sun's radiant energy is absorbed; θ is the solar zenith angle on the nucleus ($\theta=0$ at the subsolar point); r is the heliocentric distance; A_1 is the albedo in the infrared region (15–30 microns) where thermal emission would take place from the nucleus; σ is the Stefan-Boltzmann constant; T is the temperature; $Z(T)$ is the vaporization rate; and $L(T)$ is the latent heat of vaporization in calories per mole.

Equation (38) can be integrated over the surface of the nucleus to yield

$$F_0(1-A_0)\frac{S}{r^2}=4S(1-A_1)\sigma T^4+4SZ(T)L(T) \qquad (39)$$

where $S=\pi r_c^2$ is the cross-sectional area of the nucleus. If we introduce the total vaporization rate, $Q=4SZ$, then we have

$$F_0(1-A_0)\frac{S}{r^2}=4S(1-A_1)\sigma T^4+QL \qquad (40)$$

Note that the latent heat of vaporization $L(T)$ does not change significantly over temperatures of interest in comets. It is 11,700 calories per mole at 150 K and 11,220 calories per mole at 250 K. A value of 11,500 calories per mole is appropriate for comets.

The different energy regimes for a comet approaching the sun can be seen by inspection of equation (40). Far from the sun, the total vaporization rate Q is very low, and the energy absorbed from the sun is reradiated. As the comet moves closer to the sun, vaporization is still negligible and the temperature increases to balance the increased solar input. Eventually the temperature increases to a value at which vaporization begins, and both terms on the right-hand side of equation (40) are important. As the comet moves still closer, certainly by around 1 astronomical unit, all of the energy from the sun goes into vaporization, and the first term on the right-hand side is negligible. In this case, the vaporization rate varies as r^{-2} and the temperature varies slowly. Vaporization for most comets becomes important for heliocentric distances of roughly 3 astronomical units.

Accelerations in Tails

Accelerations of features in tails can be measured from a time sequence of photographs and are often expressed in terms of the parameter $(1 - \mu)$. The curvature of tails is also a function of $(1 - \mu)$. Consider a particle moving away from a comet nucleus in the tail. Clearly an extra force is operating on the particle and, by analogy with equation (2), the net gravitational force can be written as

$$F = \mu \frac{GM_\odot m}{r^2} \qquad (41)$$

The parameter μ has been introduced to indicate the effect of the extra force, assumed also to vary as the inverse square. If there is no extra force, $\mu = 1$. So the parameter of interest is the extra force, or

$$(1 - \mu) = \frac{F_{\text{extra}}}{F_{\text{gravity}}} \qquad (42)$$

For dust particles, the extra force is produced by radiation pressure. The force for clouds of plasma is not clear.

For higher values of $(1 - \mu)$, particles or features have higher accelerations into the tail. For tail shapes, lower values of $(1 - \mu)$ have a higher curvature.

Densities in Cometary Atmospheres

Densities in comas are often calculated in terms of the Haser model. Molecules are assumed to flow from the nucleus at constant speed v_0, and they are dissociated by the solar radiation field with the time for the density to fall to $1/e$ of its initial value given by τ_0. Then

$$\frac{dN}{dt} = -\frac{N}{\tau_0} \qquad (43)$$

where N is the number density. The distance traveled by the average molecule before dissociation is $R_0 = \tau_0 v_0$. Radial outflow at v_0 causes the density to drop as r^{-2} while dissociation alone would cause a drop as e^{-r/R_0}. The two processes together produce

$$N(r) = \left(\frac{R}{r}\right)^2 N(R) e^{-r/R_0} \qquad (44)$$

where R is the radius of the nucleus.

Molecules that are created from parent molecules and destroyed by photodissociation would follow a straightforward generalization of equation (44), or

$$N(r) = \left(\frac{R}{r}\right)^2 N(R)[e^{-r/R_0} - e^{-r/R_1}] \qquad (45)$$

where R_1 refers to the parent molecules and R_0 refers to the offspring molecules. Equation (45) must be integrated along the line of sight to give the column density along a line of sight that has a distance of closest approach to the nucleus of ρ. Additional changes in notation are $\beta_0 = 1/R_0$, and so on, and $\beta_0 \rho = x$. Thus

$$S(x) \propto N(\rho) \propto \frac{1}{x}\left[B(x) - B\left(\frac{\beta_1}{\beta_0}x\right)\right] \qquad (46)$$

where $S(x)$ is the surface brightness, which is proportional to the column density $N(\rho)$, and $B(x)$ is related to the modified Bessel function of zero order.

The physical significance of equation (46) is illustrated in Figure A.1, where we have plotted $\log S$ versus $\log \rho$. For simple radial outflow at constant speed, the plot would be a straight line with a slope of -1. For creation of molecules, the bright-

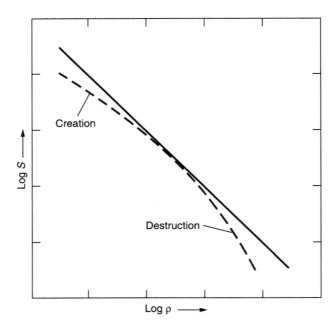

Figure A.1 Illustration of the variation in brightness for a species being created and destroyed in a comet. A species that is being neither created nor destroyed would follow the solid line. Most species are created near the nucleus and then destroyed far from the nucleus; they follow the dashed line.

ness drops more slowly; for destruction of molecules, it drops more rapidly.

The Haser model rests on certain simplifying assumptions, some of which have been mentioned above. More detailed treatments are available in the literature.

The hydrogen cloud around comets requires a similar treatment. If Q_H is the total production rate of hydrogen, v_H is the average outflow speed, and other processes are neglected, the density of hydrogen atoms n_H is given by

$$n_H = \frac{Q_H}{4\pi r^2 v_H} \qquad (47)$$

where n_H is in atoms per cubic centimeter. Hydrogen atoms are destroyed by photoionization (t_{rad}) and by charge exchange with the solar wind (t_{sw}). The total lifetime for hydrogen atoms is

$$t_H = (t_{rad}^{-1} + t_{sw}^{-1})^{-1} \qquad (48)$$

The effects of radiation pressure (from solar Lyman-alpha) are important for hydrogen atoms.

Equation (47) becomes

$$n_H(x,y,z) = \frac{Q_H e^{-t/t_H}}{4\pi v_H (x^2 + y^2 + z^2)} \qquad (49)$$

where t is the time of travel. The issue is further complicated by the fact that for a given energy for the hydrogen atom, any point in the atmosphere can be reached by two separate trajectories. An analytic expression can be obtained for $n_H(x,y,z)$ that must then be numerically integrated over an assumed distribution of velocities and along the line of sight to obtain the column density. The results are distributions displaced in the antisolar direction and symmetrical about the radius vector. The results are in reasonable agreement with observations.

B

How to Calculate
Positions of a Comet

Celestial mechanicians spend a great deal of their time trying to predict the future, and they are successful only in a limited sense. We don't know any rich celestial mechanicians who have been able to predict the course of the Dow-Jones average over the years. We do know some very talented practitioners of the art who can predict the future positions of spacecraft, comets, and planets with uncanny accuracy. Their tool in this exacting work is not a crystal ball but a state-of-the-art digital computer. As we have already described, a number of steps can be used in the process. For a newly discovered comet, the steps might include:

1. Calculation of the initial elements of the comet's orbit, given at least three observations.

2. Prediction of the comet's subsequent motion for a short time into the future. The initial elements are seldom

accurate enough to permit long-term predictions. They are used, in part, to keep track of the comet while additional position observations are made.

3. Refinement of the comet's orbital elements as more position observations are made.

4. Calculation of a definitive orbit for a longer time into the future. This process usually involves all the position observations made while the comet was visible from earth. In addition, this step involves the detailed calculation of the perturbations on the comet's motion caused by the planets and may include additional effects such as nongravitational forces. If the comet is a periodic one, this step ultimately will involve the prediction of the orbital elements for the next pass through perihelion.

From a computational point of view, the most straightforward of these steps are step 2 and the portion of step 4 in which the celestial mechanician uses the set of orbital elements to predict short-term future positions of the comet, without further considering planetary perturbations. This step can be done successfully on a home computer. In the following pages we will describe for you a simple computer program that you can implement on your home system. The program is written in the Pascal language and does not use any extensions that are found in various implementations of the language. For those of you who do not have a Pascal compiler, we will append, without comment, a BASIC program that will do the same calculations.

The Calculation Steps

The computer program uses moderately complex mathematics. In addition, the transformations between the numerous coordinate systems used in astronomy employ spherical trigonometry. It is simply not feasible here to explain all of these equations starting from first principles. Libraries have volumes on the subject. In the Suggested Readings we list some books to consult for the details. Our limited objective is to tell you what the program will accomplish for you.

The Output

To show what the computer will do as it executes the program, we will start at the end, the output routine, listed here in its entirety.

PROGRAM SECTION 1
OUTPUT

```
PROCEDURE WriteResult;

{ Write out the results calculated by the comet ephemeris
  program.  The program calculates the right ascension and
  declination to the  nearest second of time or second of arc.
  As the seconds are beyond the accuracy of the calculations,
  they are not written.                                      }

VAR
 {Hour-minute and degree-minute variables for output}
   H, M1, S1, D, M2, S2: Integer;
   SHour, SMin: Integer;   {Sidereal time hour and minute}

BEGIN   {WriteResult}
   Writeln;
   Writeln;
   Writeln('------POSITION OF COMET------');
   Writeln;
   Writeln(' Date: ', Month:2,'/',Day:2,'/',Year:4,'  Julian Day: ',
   JD : 11 : 3);
   STime := STime * RH;
   SHour := Trunc(STime);
   SMin := Trunc(60.0 * (STime - SHour));
   Write('Local Time: Solar = ', Hour : 2, ':', Min : 2); •
   Writeln('  Sidereal = ', SHour : 2, ':', SMin : 2);
   Writeln;
   Writeln('  COORDINATES OF THE SUN:  ');
   Writeln('    X = ', XSun:6:4,' Y = ',YSun:6:4,' Z = ',ZSun:6:4);
   HrToHHMMSS(RASun, H, M1, S1);
   DegToDDMMSS(DecSun, D, M2, S2);
   Declination(D, M2, S2);
   Writeln('    R.A. = ',H:2,' ',M1:2,' DEC = ',D:4,' ', M2:2);
   Writeln;
   Writeln('  COORDINATES OF THE COMET ');
   Writeln('    X = ', XBody:6:4,' Y = ',YBody:6:4,' Z = ',ZBody:6:4);
   HrToHHMMSS(RABody, H, M1, S1);
   DegToDDMMSS(DecBody, D, M2, S2);
   Declination(D, M2, S2);
   Writeln('    R.A. = ',H:2,' ' M1:2,' DEC = ',D:4,' ',M2:2);
   Writeln('    R = ', DistSB : 4 : 2, ' Delta = ', DistEB : 4 : 2);
   Writeln('    Elongation of Body = ', Theta : 5 : 1);
   HrToHHMMSS(HA, H, M1, S1);
   DegToDDMMSS(Az, D, M2, S2);
   X := 0.0;
   IF HA > 12 THEN
      BEGIN
         X := 1.0;
         HA := 24 - HA
      END;
   Write('            Hour Angle = ', H : 4, '  ', M1 : 2);
```

```
    IF X = 0.0 THEN Writeln(' West')
       ELSE Writeln(' East');
    Writeln('            Azimuth = ', D : 3, ' ', M2 : 2);
    Elev := Elev * RD;
    D := Trunc(Elev);
    M2 := Abs(Trunc(60.0 * (Elev - D)));
       Writeln('            Elevation = ', D : 3, ' ', M2 : 2);
END; {WriteResult}
```

The program is designed to calculate and write out the position of a comet (it can be used for asteroids and planets as well) at some time on a given date. The position is given in several coordinate systems. The first coordinates—called XBody, YBody, and ZBody in the program and X, Y, and Z in the output—are the comet's position in a three-dimensional, rectangular coordinate system referred to as the heliocentric equatorial coordinate system, which places the comet in space. In this right-handed coordinate system, the plane of the earth's equator defines the xy plane, with the x axis in the direction of the vernal equinox and the z axis toward the north pole. The X, Y, and Z coordinates of the sun printed by the output routine are in the geocentric equatorial coordinate system, which differs from the heliocentric system only in that the origin of the system is shifted from the sun to the earth. (Within the program the heliocentric ecliptic is used. In that system the XY plane is the plane of the ecliptic, which makes an angle of about 23°27′—called the obliquity of the ecliptic—with the equator.)

The program writes out the right ascension and declination of the comet, to help place it among the stars. Right ascension and declination are coordinates on the sky analogous to longitude and latitude on the earth, and right ascension is measured from the vernal equinox. The program next prints out the comet's distance from the sun, R; its distance from the earth, delta; and its apparent angular distance from the sun, called its elongation (theta in the program). The two distances can be used to calculate the comet's magnitude—which we will not consider here—and they help us decide if it is visible.

The program finally prints out coordinates that help an observer standing on the earth's surface to locate the comet in the sky. These coordinates are the hour angle (HA), altitude, and azimuth. Hour angle is measured similarly to right ascension, except that the zero point for hour angle is the point on the sky where the celestial equator crosses the meridian. While the right ascension of an object is fixed in time (or, more precisely, changes very slowly because of precession), the hour angle of an object changes with sidereal time. The altitude is the angular distance of an object above the horizon, and its azimuth is measured around the horizon. The program maintains all angles in radians in its internal calculations and converts to degrees or hours when it writes answers to output.

To calculate the 11 parameters described above, the program must calculate and use a number of intermediate quantities. Of these, only the coordinates of the sun are actually printed out. We will describe the intermediate quantities as we proceed.

The Main Program

The main program, which we call EPHEMERIS, is mostly contained inside a large loop that calculates positions of the comet for a series of dates.

PROGRAM SECTION 2
MAIN PROGRAM

```
PROGRAM EPHEMERIS (Input, Output);

CONST    {global}
   PI = 3.14159265359;
   TwoPi = 6.28318530718;
   RD = 57.295779513;           {degrees in a radian}
   RH = 3.81971863;             {hours in a radian}
   TAU = 5.77557E-03;           {days for light to travel 1 AU}
   JD1900 = 2.4150205E+06;      {Julian date on epoch 1900, Jan 0.5}
   NewStyle = true;             {we will use Gregorian calendar only}

VAR
 {orbital Elements and their epoch}
   E, LongNode, LongPeri, Incl, A, P, PeriDay, N: Real;
 {Sun-Earth, Sun-Body and Earth-Body Distance}
   DistSE, DistSB, DistEB     : Real;
 {elongation: Sun-Body angle seen from Earth}
   Theta                      : Real;
 {equatorial Cartesian coordinates in astronomical units}
   X, Y, Z                    : Real;   {geocentric coordinates}
   XSun, YSun, ZSun: Real;              {geocentric coordinates of sun}
   XBody, YBody, ZBody        : Real;   {heliocentric coordinates}
   Epoch                      : Integer; {epoch of orbital elements}
   JE                         : Real;   {Julian day of 1/1/Epoch}
   Obliq                      : Real;   {obliquity of the ecliptic}
```

```
     Zeta0, Zee, Th                   : Real;      {precession constants}
     EA, M                            : Real;      {eccentric and mean anomaly}
     PX, PY, PZ, QX, QY, QZ           : Real;      {coordinate rotation vectors}
     RA1, Dec1                        : Real;      {equatorial sky coordinates}
     RABody, DecBody                  : Real;      {equatorial sky coordinates}
     RASun, DecSun                    : Real;      {equatorial sky coordinates}
     Lat                              : Real;      {latitude}
     Elev, Az, HA                     : Real;      {horizon-system coordinates}
     Month, Day, Year, Hour, Min: Integer;        {date/time}
     JD, JH                           : Real;      {Julian date/fractional day}
     STime                            : Real;      {sidereal time}
     Key                              : Char;

{ place PROCEDURE/FUNCTION      Arctg2                    here
                                Julian
                                InputElements
                                Kepler
                                Sun
                                SidTime
                                Precession
                                PrecessElements
                                SetPQ
                                CalcXYZ
                                RADec
                                HrToHHMMSS
                                DegToDDMMSS
                                Declination
                                SkyCoordinates
                                Iteration
                                Elongation
                                Write Results             }

{ ---------------------------------------------------------- }

BEGIN   {Ephemeris}
     Writeln('Input your latitude in decimal degrees:');
     Writeln('  Negative for the southern hemisphere.');
     Readln(Lat);
     Lat := Lat / RD;
     InputElements;
     SetPQ;
     JE := Julian(1, 1, Epoch, NewStyle);
     Obliq := Obliquity(JE);

REPEAT
     Writeln('Input date of interest: Month  Day  Year ');
     Writeln('  Inputs are integers');
     Readln(Month, Day, Year);
     IF Year < 100 THEN Year := Year + 1900;
     JD := Julian(Month, Day, Year, NewStyle);
     PrecessElements(JE, JD, Zeta0, Zee, Th);
     Writeln('Input time of interest: HH  MM');
     Readln(hour, Min);
     JH := (Hour + Min / 60.0) / 24.0;
     JD := JD + JH - 0.5;
     SunJD, JE, XSun, YSun, ZSun, DistSE);
     RADec(XSun, YSun, ZSun, DistSE, RASun, DecSun);
     M := (JD - PeriDay) * N;
     IF M < 0.0 THEN
     M := M + TwoPi;
     Iteration(M, X, Y, Z);

     {planetary aberration}
     M := M - TAU * DistEB * N;
```

```
    Iteration(M, X, Y, Z);
    RADec(X, Y, Z, DistEB, RABody, DecBody);  {Equinox of Epoch}

    {Precess to equinox of date}
    RA1 := RABody;
    Dec1 := DecBody;
    Precession(Zeta0, Zee, Th, RA1, Dec1, RABody, DecBody);
    Theta := Elongation(RABody, DecBody, RASun, DecSun);
    STime := SidTime(JD, JH);
    SkyCoordinates(RABody, DecBody, STime, HA, Az, Elev);
    WriteResult;
    Writeln;
    Writeln('  Do you want another Calculation? (Y/N): ');
    Read(Key)
UNTIL (Key = Chr(78)) or (Key = Chr(110))

END. {Ephemeris}
```

The Inputs

The observer's latitude and the orbital elements are assumed
to be fixed quantities and are input outside the REPEAT loop.
The routine InputElements initializes the orbital elements. The
program section shown here does not conform to sound pro-
gramming practice; you would have to recompile the program
each time it was run with a new set of elements. We will leave
it up to you to decide how you want to input elements for the
comet or other celestial object and implement the necessary
input routine. You might build a sequential data file of orbital
elements of comets that are expected to pass perihelion in the
near future, then read in the elements for any comet of interest. The elements input by procedure InputElements here were
valid for comet Halley at the time of the latest perihelion. We
list a few results to help you test your program; see page 249.

PROGRAM SECTION 3
INPUT ORBITAL ELEMENTS

```
PROCEDURE InputElements;

BEGIN {InputElements}
    E := 0.967276;
    LongNode := 1.01482798;
    LongPeri := 1.95211742;
    Incl := 2.83160961;
    A := 17.941104;
    P := 75.99303;
    PeriDay := 2446470.45174;
    Epoch := 1986;
    N := TwoPi / (P * 365.2422)
END; {InputElements}
```

After the orbital elements are entered in some way, and before the input of the first date, the main program calls procedure SetPQ to calculate the two vectors (Px, Py, Pz and Qx, Qy, Qz) that are used to transform coordinates from a two-dimensional system in the plane of the orbit to the three-dimensional heliocentric ecliptic system.

PROGRAM SECTION 4
SET COORDINATE ROTATION VECTORS

```
PROCEDURE SetPQ;

{  The vectors P and Q will be used to rotate the coordinate
   system from the orbit plane to the heliocentric ecliptic system. }

VAR
    CLP, SLP, CLN, SLN, CI, SI: Real;

BEGIN  {SetPQ}
    CLP := Cos(LongPeri);   {argument of perihelion}
    CLN := Cos(LongNode);   {longitude of ascending node}
    SLP := Sin(LongPeri);
    SLN := Sin(LongNode);
    SI  := Sin(Incl);       {inclination of orbit to ecliptic}
    CI  := Cos(Incl);
    PX  := CLP * CLN - SLP * SLN * CI;
    PY  := CLP * SLN + SLP * CLN * CI;
    PZ  := SLP * SI;
    QX  := -SLP * CLN - CLP * SLN * CI;
    QY  := -SLP * SLN + CLP * CLN * CI;
    QZ  := CLP * SI
END;   {SetPQ}
```

The vectors can be calculated once and for all, unless new orbital elements are adopted. It is important to note that the parameter called LongPeri is the argument of perihelion, the definition of which differs from the longitude of perihelion. The necessary equations are:

$$
\begin{aligned}
Px &= \cos \omega \cos \Omega - \sin \omega \sin \Omega \cos i \\
Py &= \cos \omega \sin \Omega + \sin \omega \sin \Omega \sin i \\
Pz &= \sin \omega \sin i \\
Qx &= -\sin \omega \cos \Omega - \cos \omega \sin \Omega \cos i \\
Qy &= -\sin \omega \sin \Omega + \cos \omega \cos \Omega \cos i \\
Qz &= \cos \omega \sin i
\end{aligned}
$$

Finally, the date and time for which calculations are desired are input, and the program continues to request additional input dates until you request it to stop. Generally the program will use the interval in days between certain dates. Therefore, it is most convenient to use the Julian date, which is the number of days that have elapsed since noon Greenwich mean time on January 1, 4713 B.C. Since the Julian day begins at noon and the solar day begins at midnight, the Julian date at the beginning of a certain solar day is an integer number plus 0.5. For example, January 1, 1992, is Julian date 2448622.5. The function Obliquity calculates the obliquity of the ecliptic for the same epoch as the orbital elements.

PROGRAM SECTION 5
CALCULATE JULIAN DATE

```
FUNCTION Julian (Month, Day, Year: Integer; NewStyle: Boolean): Real;

{ Calculates the Julian day number.  NewStyle is false for the
  Julian calendar and true for the Gregorian calendar.          }

VAR
     A1, B, C, T, U: Real;
     X: Integer;

BEGIN  {Julian}
   IF Month <= 2 THEN X := 1
     ELSE X := 0;
   IF NOT NewStyle THEN C := 2.0
     ELSE C := (Year - X) div 100;
   T := 365.25 * (Year - X);
   U := Trunc(T / 1000.0) * 1000.0;
   T := T - U;
   A1 := Trunc(Trunc(T) - 0.75 * C);
   B := Trunc(367.0 * ((Month - 2.0) / 12.0 + X));
   IF B < 0.0 THEN B := B + 1.0;
   Julian := 1.7210885E+6 + Day + U + A1 + B
END;  {Julian}


FUNCTION Obliquity (JD: Real): Real;

{ Calculate obliquity of the ecliptic for Julian day, JD.
  Unit is radians.                                         }

CONST
   Ob0 = 0.409319755;           {obliquity for 1900.0}
   Ob1 = 2.27135E-04;           {rate of change of obliquity}
   JC = 3.6525E+04;             {days in a Julian Century}
   JD0 = 2.4150005E+06;         {Julian day on 1/1/1900 = 1900.0}

BEGIN  {Obliquity}
     Obliquity := Ob0 - Ob1 * (JD - JD0) / JC
END;   {Obliquity}
```

The Calculation Loop

The first calculation step within our main loop is to compute the position of the sun. The procedure Sun calculates the geocentric equatorial coordinates of the sun and the earth-sun distance, given two Julian dates—the date of the desired position, JD, and the date JE (or epoch) of the equinox to which the position is referred. This latter factor involves the precession of the equinoxes, which we will again leave to additional reading. The procedure to calculate the sun's position contains an approximation that is sufficiently accurate for our purposes.

PROGRAM SECTION 6
POSITION OF THE SUN

```
PROCEDURE Sun (JD, JE: Real; VAR XSun, YSUN, ZSUN, DistSE: Real);

{  Calculates the geocentric equatorial coordinates of the
   Sun, and the Earth-Sun distance in astronomical units, for
   the Julian Date JD referred to the equinox on Julian Date JE.}

CONST
    JD2 = 2.451545E+6;      {epoch 2000.0}

VAR
    DeltaD: Real;           {interval in days}
    G: Real;                {mean anomaly}
    L: Real;                {mean longitude}
    EclLong: Real;          {ecliptic Longitude of Sun}
    X: Real;                {working variable}

BEGIN   {Sun}
    DeltaD := JD - JD2;
    G := (357.528 + 0.9856003 * DeltaD);
    WHILE G < 0.0 DO G := G + 360.0;
    G := G / RD;
    L := (280.460 + 0.9856474 * DeltaD);
    WHILE L < 0.0 DO L := L + 360.0;
    L := L / RD;
    EclLong := L + 0.03344 * Sin(G) + 3.49E-4 * Sin(2.0 * G);
{Precess to Epoch JE}
    EclLong := EclLong + 2.437E-4 * (JD - JE) / 365.25;
    DistSE := 1.00014 - 0.01671 * Cos(G) - 0.00014 * Cos(2.0 * G);
    XSun := DistSE * Cos(EclLong);
    YSun := DistSE * Sin(EclLong) * Cos(Obliq);
    ZSun :=.DistSE * Sin(EclLong) * Sin(Obliq)
END;    {Sun}
```

The procedure RADec converts the geocentric equatorial coordinates to right ascension and declination. The procedure needs a function Arctg2, which is a two-argument arctangent that yields a result in the proper quadrant.

PROGRAM SECTION 7
CONVERT GEOCENTRIC EQUATORIAL
COORDINATES TO RIGHT ASCENSION
AND DECLINATION

```
FUNCTION Arctg2 (Sn, Cs: Real): Real;

{  Calculates the arctangent of Sn/Cs in the proper quadrant. }

VAR
    XT: Real;

BEGIN    {Arctg2}
    IF Abs(Cs) < 1.0E-08 THEN
        BEGIN
            IF Sn < 0 THEN XT := 0.5 * Pi
                ELSE XT := 1.5 * Pi
        END
    ELSE
        BEGIN
            XT := Arctan(Sn / Cs);
            IF Cs < 0 THEN XT := XT + Pi
            ELSE IF Sn < 0 THEN XT := XT + 2.0 * Pi
        END;
    Arctg2 := XT
END;    {Arctg2}

PROCEDURE RADec (XG, YG, ZG, Distance: Real; VAR RARad, DecRad: Real);

{  Convert equatorial geocentric  coordinates XG, YG and ZG to
   Right Ascension, RARAD, and Declination, DecRad,  in radians.
   Distance is the distance from the Earth to the body.}

VAR
    SA, CA: Real;

BEGIN  {RADec}
    SA := ZG / Distance;
    CA := Sqrt(1.0 - Sqr(SA));
    DecRad := Arctg2(SA, CA);
    SA := YG / (Distance * CA);
    CA := XG / (Distance * CA);
    RARad := Arctg2(SA, CA)
END;    {RADec}
```

The program next calculates a quantity called the mean anomaly, *M*, and calls the routine Iteration, which does one of the important jobs of the program. Iteration in turn calls procedure Kepler and procedure CalcXYZ to execute its function.

PROGRAM SECTION 8
ROUTINES TO CALCULATE GEOCENTRIC EQUATORIAL COORDINATES

```
PROCEDURE Kepler (Eccen: Real; VAR M, EA: Real);

{  Solves Kepler's equation iteratively.  The variables are
   Eccen = eccentricity, M = mean anomaly, EA = eccentricanomaly.}

CONST
   Eps = 1.0E-08;

VAR
   X, EB: Real;

BEGIN {KEPLER}
 {Make sure 0<=M<TwoPi}
   X := M / TwoPi;
   M := (X - Trunc(X)) * TwoPi;
   EB := M;
   REPEAT
      EA := EB;
      EB := EA - (EA - M - Eccen * Sin(EA)) / (1.0 - Eccen * Cos(EA))
      UNTIL Abs(EB - EA) / EB < EPS;
      EA := EB
   END;  {Kepler}

PROCEDURE CalcXYZ (A, E, EA: Real; VAR Xeq, Yeq, Zeq: Real);

{  Procedure uses semi-major axis, A, eccentricity, E, and  eccentric
   anomaly, EA, to calculate coordinates of a celestial body.           }

VAR
   X1, Y1, Xecl, Yecl, Zecl: Real;

{Obliq, PX, PY, PZ, QX, QY, QZ are GLOBAL variables.  }

BEGIN {CalcXYZ}

{  Calculate cartesian coordinates of position in orbit, where
   the x-axis is toward perihelion.                         }

   X1 := A * (Cos(EA) - E);
   Y1 := A * Sqrt(1.0 - E * E) * Sin(EA);

{  Transform to ecliptic heliocentric coordinates, where the
   x-axis is toward the Vernal equinox.                    }

   Xecl := X1 * PX + Y1 * QX;
   Yecl := X1 * PY + Y1 * QY;
   Zecl := X1 * PZ + Y1 * QZ;

{  Transform to equatorial heliocentric coordinates, x-axis same as
   above.                                                  }

   Xeq := Xecl;
   Yeq := Yecl * Cos(Obliq) - Zecl * Sin(Obliq);
   Zeq := Yecl * Sin(Obliq) + Zecl * Cos(Obliq)
END;  {CalcXYZ}
```

```
PROCEDURE Iteration (VAR M, XG, YG, ZG: Real);

{ Procedure invokes solution of Kepler's equation, finds the
  equatorial  heliocentric coordinates, XBody..., of a body,
  then  translates them to equatorial geocentric coordinates,
  XG...  In the process it finds the Sun-Body and Earth-Body
  distances DistSB and Dist B.                                }

BEGIN {Iteration}
   Kepler(E, M, EA);
   CalcXYZ(A, E, EA, XBody, YBody, ZBody);
   DistSB := Sqrt(Sqr(XBody) + Sqr(YBody) + Sqr(ZBody));
   XG := XBody + XSun;
   YG := YBody + YSun;
   ZG := ZBody + ZSun;
   DistEB := Sqrt(Sqr(XG) + Sqr(YG) + Sqr(ZG))
END;  {Iteration}
```

To understand what is going on, please refer to Figure
B.1, which illustrates an elliptical orbit with the sun at one focus, labeled *S*. Around the ellipse is circumscribed a concentric
circle with radius equal to the semimajor axis of the ellipse. In
polar coordinates centered at the sun, the position of the comet
is given by angle ν—called the true anomaly—and radius *r*. We
now draw a perpendicular line from the semimajor axis of the
elliptical orbit through the comet and extend the line until it
intersects the circle at point *C*. The angle *E* at the center of
the circle as illustrated in Figure B.1 is called the eccentric
anomaly and is given by *Kepler's equation:*

$$E - e\,\sin(E) = M = n(t - T)$$

where *E* is the eccentric anomaly, *e* is the eccentricity of the
orbit, and *M* is the mean anomaly. The mean anomaly is calculated from the mean motion *n*, which is the average number
of radians the comet moves in one day; *t* is the time for which
the calculation is desired, and *T* is the time of perihelion passage of the comet. Here we see the usefulness of expressing *t*
and *T* in Julian days; then the difference is expressed in days.
One of the major problems of celestial mechanics has been to
find methods to solve this transcendental equation for *E*. Before the advent of modern computers, methods using series
expansions were employed. The procedure Kepler uses a more
direct iterative scheme. The scheme is stable and converges for
values of *e* less than 0.98 or 0.99—which means that it will
converge for all the periodic comets that have been seen at
more than one appearance near the sun, including comet Halley with $e = 0.967$ . . . .

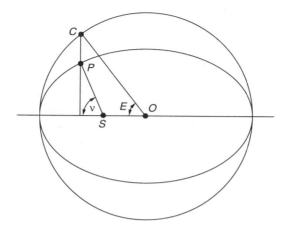

Figure B.1 The definition of certain quantities used in calculating the ephemeris of a body moving in an elliptical orbit. A concentric circle, with diameter equal to the major axis of the ellipse, is circumscribed around the elliptical orbit. The label O marks the centers of the shapes. The focus containing the sun is labeled S and the comet or planet is at the point labeled P. A vertical line perpendicular to the major axis of the ellipse is drawn through P. The line intersects the circle at C. The angle E is the eccentric anomaly and v is the true anomaly.

Once the eccentric anomaly has been found, the coordinates of the comet in its orbit follow from the equations

$$x1 = r \sin v = a \; \text{sqrt}(1 - e{*}e) \sin E$$
$$y1 = r \cos v = a \; (\cos E - e)$$

where a is the semimajor axis of the elliptical orbit. The coordinate system is a rectangular system with the x axis toward perihelion. To convert these coordinates to the three-dimensional ecliptic system, with the x axis toward the vernal equinox, the vectors Px, Py, Pz, Qx, Qy, and Qz described above are used in the equations (in the program the coordinates are called xecl, yecl and zecl):

$$xecl = x1 \; Px + y1 \; Qx$$
$$yecl = x1 \; Py + y1 \; Qy$$
$$zecl = x1 \; Pz + y1 \; Qz$$

These coordinates are then rotated to the equatorial coordinates (called xeq, yeq, and zeq) by the equations:

$$xeq = xecl$$
$$yeq = yecl \cos \epsilon - zecl \sin \epsilon$$
$$zeq = yecl \sin \epsilon + zecl \cos \epsilon$$

The procedure Iteration first calls the procedure Kepler to solve Kepler's equation, then calls CalcXYZ, which first finds the coordinates x1 and y1 in the orbit, transforms that result to heliocentric ecliptic coordinates and then to heliocentric equatorial coordinates, and returns the latter coordinates. Finally, procedure Iteration finds the geocentric equatorial coordinates of the comet, which it returns to the main program.

We now have a position of the comet centered at the earth. As seen from earth, however, this is not the position seen at time t; it is the position seen at time t plus the time it takes light to reach us from the comet. The position of a fast-moving object such as a comet can change significantly during the time light takes to reach us. Therefore, we correct the mean anomaly for the light time and redo the Iteration. The program next corrects the position of the comet for precession of the equinoxes.

The remainder of the main program calculates the position of the comet in other coordinate systems and outputs the result.

PROGRAM SECTION 9
REMAINING CALCULATIONS

```
PROCEDURE PrecessElements(JD1, JD2 : Real; VAR Zeta0, Z, Theta :Real);

{       Fully general equatorial precessional elements, to precess from JD1 to
        JD2.}

CONST
    JD0       = 2.4150005E+06;    {JD on 1900.0}
    Cent      = 3.652422E+04;     {Days in a tropical century}
    Zeta1     = 1.1171319E-02;    {Constants to calculate parameter}
    Zeta11    = 6.7680E-06;
    Zeta2     = 1.464E-06;
    Zeta3     = 8.272E-08;
    Z2        = 3.835E-06;        {Constant to calculate z parameter}
    Theta1    = 9.7189726E-03;    {Constants to calculate parameter}
    Theta11   = 4.135E-06;
    Theta2    = 2.065E-06;
    Theta3    = 2.036E-07;


VAR
    T, T0, T2, T3 : Real;    {Time variables}
```

```
BEGIN   {Precession Elements}
   T0    :=  (JD1 - JD0)/Cent;   {Tropical centuries since 1900.0}
   T     :=  (JD2 - JD1)/Cent;   {Tropical centuries since JD1}
   T2    :=  T * T;
   T3    :=  T * T * T;
   Zeta0 := (Zeta1 + Zeta11 * T0) * T + Zeta2 * T2 +Zeta3 * T3;
   Z     := Zeta0 + Z2 * T * T;
   Theta := (Theta1 - Theta11 * T0) * T -Theta2 * T2 - Theta3 * T3
 END;    {Precession Elements}

PROCEDURE Precession(Zeta0, Z, Theta, RA0, Dec0 : Real;
         VAR RA1, Dec1 : Real);

{General precession of the equinoxes program.}

VAR

  CD, SD, CR, SR, CT, ST, CR1, SR1  : Real;
  A, B, C, SA, CA         : Real;

BEGIN   {Precession}

{Required trig functions of known values}
  CD := Cos(Dec0);
  SD := Sin(Dec0);
  CR := Cos(RA0 + Zeta0);
  SR := Sin(RA0 + Zeta0);
  CT := Cos(Theta);
  ST := Sin(Theta);
{Precession equations}
  A := CD * SR;                 { Cos(Dec1) * Sin(RA1 - Z) }
  B := CT * CD * CR - ST * SD;  { Cos(Dec1) * Cos(RA1 - z) }
  C := CT * SD + ST * CD * CR;  { Sin(Dec1)                }
{Solve for RA1 and Dec1}
  CR1  := Sqrt(1.0 - Sqr(C));                {Abs(Cos(Dec1))}
   SA := A / CR1;
   CA := B / CR1;
   RA1 := Arctg2(SA, CA);                    {actually: RA1 - Z}
   IF RA1 < 0.0 THEN RA1 := RA1 + 2.0 * Pi;
   SR1 := Sin(RA1);                          {actually Sin(RA1 - Z)}
   RA1 := RA1 + Z;                           {result: RA1}
   SA := C;
   CA := A / SR1;
   Dec1 := Arctg2(SA, CA)                    {result: Dec1}
END;    {Precession}


FUNCTION SidTime (JD, JH: Real): Real;    {time in radians}

{ Calculates the sidereal time for a given Julian day, JD,
  and fraction of a day, JH. }

VAR
   T1, X1: Real;

BEGIN    {SidTime}
   T1 := (JD - JD1900) / 36525.0;          {centuries since 1900 Jan 0.5}
   X1 := (18.64606 + 2400.0513 * T1) / 24.0 + 0.5 + JH;
   SidTime := (X1 - Trunc(X1)) * TwoPi
END;     {SidTime}
```

```
PROCEDURE SkyCoordinates (RARad, DecRad, SidTime: Real; VAR HourAngle,
        Azimuth, Elevation: Real);

{  Calculate hour angle, azimuth and elevation of a body at a
   given  sidereal time. Latitude (Lat) and PI are global
   variables.}

VAR
   SA, CA, X, Y, Z: Real;
BEGIN  {SkyCoordinates}
   HourAngle := SidTime - RARad;
   IF HourAngle < 0 THEN
      HourAngle := HourAngle + 2.0 * PI;
   X := -Cos(DecRad) * Sin(HourAngle);
   Y := Sin(DecRad) * Cos(Lat) - Cos(DecRad) * Cos(HourAngle) *
        Sin(Lat);
   Z := Sin(DecRad) * Sin(Lat) + Cos(DecRad) * Cos(HourAngle) *
        Cos(Lat);
   SA := Z;
   CA := Sqrt(1.0 - Sqr(SA));
   Elevation := Arctg2(SA, CA);
   IF Elevation > 4.72 THEN
      Elevation := Elevation - 2.0 * PI;
   SA := X / CA;
   CA := Y / CA;
   AZ := Arctg2(SA, CA)
END;    {Sky Coordinates}


PROCEDURE Declination (VAR Deg, Min, Sec: Integer);

{  Convert declinations in the range 270 to 360 degrees to
   negative declinations, with the negative sign on  the degree
   part only.                                                  }

BEGIN     {Declination}
   IF Deg > 90 THEN
      BEGIN
         Deg := Deg - 359;
         Min := 59 - Min;
         Sec := 60 - Sec
      END
END;      {Declination}


PROCEDURE DegToDDMMSS (DegRad: Real; VAR Degrees, ArcMin, ArcSec:
           Integer);

{  Convert angular coordinate in radians to degrees:minutes:seconds.}

VAR
   X: Real;   {temporary variable}

BEGIN  {DegToDDMMSS}
   DegRad := DegRad * RD;
   Degrees := Trunc(DegRad);
   X := 60.0 * (DegRad - Degrees);
   ArcMin := Trunc(X);
   ArcSec := Trunc(60.0 * (X - ArcMin))
END; {DegToDDMMSS}


PROCEDURE HrToHHMMSS (HrRad: Real; VAR Hour, TimeMin, TimeSec: Integer);
```

```
{  Convert time coordinate in radians to hours:minutes:seconds.}

VAR
   X: Real;       {temporary variable}

BEGIN   {HrToHHMMSS}
   HrRad := HrRad * RH;
   Hour := Trunc(HrRad);
   X = 60.0 * (HrRad - Hour);
   TimeMin := Trunc(X);
   TimeSec := Trunc(60.0 * (X - TimeMin))
   END;    {HrToHHMMSS}
```

Plotting the Positions of Comets

A program that plots the calculated positions of a comet on a map of the stars can be quite impressive. Such a plot might show the relevant portion of the sky with the proper orientation for your latitude and the time of the calculation, with the horizon drawn in and compass directions shown. A program

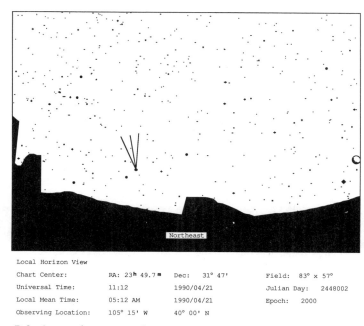

```
Local Horizon View

Chart Center:         RA: 23h 49.7m   Dec:    31° 47'    Field:  83° x 57°

Universal Time:       11:12           1990/04/21         Julian Day:  2448002

Local Mean Time:      05:12 AM        1990/04/21         Epoch:  2000

Observing Location:   105° 15' W      40° 00' N
```

Figure B.2 A sample output of a computer program to display the positions of comets. The figure shows comet Austin on April 21, 1990 (from Denver, Colorado), as calculated by VOYAGER, the Interactive Desktop Planetarium™. Besides the comet, the figure shows the horizon, Venus and the moon at lower right, and the stars with different symbols to simulate their brightness.

will make reasonable plots even if it has built in the positions of only the brightest thousand or so stars in the entire sky. Figure B.2 is a plot of the position of comet Austin made by a program in use at the University of Colorado.

BASIC PROGRAM

```
5 DEFDBL A-Z
10 REM *** INPUT NUMERICAL CONSTANTS ***
20 REM *** OB = OBLIQUITY
30 Pi = 3.14159265#
32  OB = .40914#
34  RD = 57.2957795#
36  TwoPi = 2# * Pi
40 REM *** INPUT ORBITAL ELEMENTS--ANGLES IN RADIANS***
50 REM E=ECCCENTRICITY; O1=NODE LONGITUDE; O2=PERIHELION LONGITUDE
60 REM IN=INCLINATION; O1=NODE LONGITUDE;O2=PERIHELION LONGITUDE
70 REM PD=JULIAN DAY OF PERIHELION; PH=ADDITIONAL DAY FRACTION
80 E = .967276#
82 O1=1.01482798#
84 O2=1.95211743#
90 IN = 2.83160961#
92 A=17.941104#
94 P=75.99303#
100 PD = 2446470.5#
102 PH=.45174#
110 GOSUB 6000
120 N = TwoPi / P / 365.2422#: REM MEAN MOTION
121 CLS : LOCATE 1, 8: PRINT "THIS PROGRAM CALCULATES INFORMATION"
122 PRINT "OF INTEREST FOR COMET HALLEY FOR ANY"
123 PRINT "LOCATION ON EARTH . THE OPERATION"
124 PRINT "IS SELF EXPLANATORY. ": PRINT
125 PRINT "SOME DEFINITIONS": PRINT "X, Y, Z ARE COORDINATES IN EQUATORIAL"
126 PRINT "SYSTEM IN AU.  R IS DISTANCE FROM SUN."
127 PRINT "DELTA IS DISTANCE FROM EARTH.": PRINT
128 PRINT "INPUT YOUR LATITUDE IN DECIMAL DEGREES": INPUT " NEGATIVE IF IN
SOUTHERN HEMISPHERE ";LA
129 LA = LA / RD
130 INPUT "DATE OF INTEREST? (MM,DD,YYYY) ";MM,DD,YY:S = 1: GOSUB 5000
140 INPUT "TIME OF INTEREST? (HH,MM) ";HH,M1
150 JH=(HH + M1 / 60#) / 24#
160 M=((JD-PD) + (JH - PH)) * N
165 GOSUB 7500: REM FIND POSITION OF SUN
170 KY = 0
180 GOSUB 7000: REM SOLVE KEPLER'S EQUATION
190 GOSUB 8000: REM FIND X,Y,Z OF COMET
195 REM SC=SUN-COMET DISTANCE
200 SC =SQR ( XC * XC + YC * YC + ZC * ZC)
210 X = XC + XS:Y = YC + YS:Z = ZC + ZS
220 REM CALCULATE EARTH—COMET DISTANCE
230 REM AC=RIGHT ASCENSION OF COMET; DC=DECLINATION OF COMET
240 EC = SQR (X * X + Y * Y + Z * Z)
245 IF KY = 1 THEN 250
248 DM = .005772# * EC * N:M = M-DM:KY = 1: GOTO 180
250 SA = Z / EC:CA = SQR (1#-SA * SA): GOSUB 8500
260 DC = A3:SA = Y / (EC * CA):CA = X / (EC * CA): GOSUB 8500
270 AC = A3
280 REM TH=ANGULAR SEPARATION SUN—COMET
290 CA = SIN (DC) * SIN (DS) + COS (AS1-AC) * COS (DC) * COS (DS)
300 SA=SQR (1#-CA * CA): GOSUB 8500
```

```
310 TH =A3
320 GOSUB 9000: REM FIND SIDEREAL TIME
330 HA= ST-AC: REM HA=HOUR ANGLE
332 IF HA < 0 THEN HA= HA + TwoPi
340 X = - COS ( DC ) * SIN(HA)
350 Y = SIN (DC) * COS (LA) - COS (DC) * COS (HA) * SIN (LA)
360 Z = SIN (DC) * SIN (LA) + COS (DC) * COS (HA) * COS (LA)
370 SA = Z:CA=SQR (1#-SA * SA): GOSUB 8500
380 EL = A3: SA= X / CA:CA=Y / CA: GOSUB 8500
385 IF EL > 4.72# THEN EL = EL-TwoPi
390 AZ = A3
400 IF JD > 2446470.5# THEN 430
410 MT= 5.47 + (5! * LOG (EC) + 11.1 * LOG (SC)) / 2.3026
420 GOTO 440
430 MT = 4.94 + (5! * LOG (EC) + 7.68 * LOG (SC)) / 2.3026
440 MN = 14.1 + (5! * LOG (EC) + 5! * LOG (SC)) / 2.3026
2000 CLS : PRINT " POSITION OF COMET HALLEY ": PRINT
2010 PRINT "DATE: ";MM;"/";DD;"/";YY;" JULIAN DAY ";JD: PRINT
2020 RA = 3.81971863#: REM CONVERT RADIANS TO HOURS
2022 ST = ST * RA:H = INT (ST):M2 = INT (60 * (ST-H))
2024 PRINT "LOCAL TIME: SOLAR= ";HH;" ";M1;" SIDEREAL= ";H;" ";M2
2026 PRINT
2028 PRINT "COORDINATES OF THE SUN: "
2029 XS= INT (10000 * XS) / 10000:YS = INT (10000 * YS) / 10000:ZS= INT(10000
* ZS) / 10000
2030 PRINT "        X= ";XS;" Y= ";YS;" Z= ";ZS
2040 AS1 = AS1 * RA : H = INT (AS1) : M2 = INT(60 * (AS1-H))
2050 DS = DS * RD:D = INT (DS): M3 = INT (60 * (DS-D)): IF D <= 90 THEN 2070
2060 D = D-359:M3 = 60-M3
2070 PRINT "        R.A.= ";H;" ";M2;" DEC= ";D;" ";M3
2072 PRINT : PRINT "COORDINATES OF THE COMET: "
2078 XC = INT ( 10000 * XC) / 10000:YC = INT ( 10000 * YC)/10000: ZC=INT(10000
* ZC) / 10000
2080 PRINT "        X= ";XC;"  Y= ";YC;"  Z= ";ZC
2090 AC= AC * RA:H= INT(AC):M2 = INT (60 * (AC-H))
2100 DC=DC * RD:D =INT (DC):M3= INT (60 * (DC-D)): IF D < = 90 THEN 2120
2110 D=D-359:M3=60-M3
2120 PRINT "        R.A.=";H;" ";M2;" DEC= ";D;" ";M3
2130 PRINT "        R= "; INT (100 * SC + .5) / 100;
2140 PRINT " DELTA= "; INT (100 * EC + .5) / 100
2145 TH= INT (100 * TH * RD + .5) / 100
2150 PRINT "        SUN-COMET ANGLE= ";TH
2160 HA = HA * RA:X = 0: IF HA > 12 THEN X = 1:HA = 24 - HA
2170 H = INT (HA):   M2 = INT (60 * (HA-H))
2180 PRINT "        HOUR ANGLE= ";H;" ";M2;
2190 IF X=0 THEN PRINT " WEST"
2200 IF X=1 THEN PRINT " EAST"
2210 AZ = AZ * RD:D = INT(AZ):M3 = INT (60 * (AZ-D))
2220 PRINT "        AZIMUTH= ";D;" ";M3
2230 EL=EL * RD:D = INT (EL):M3 = INT (60 * (EL-D))
2240 PRINT "        ELEVATION' ";D;" ";M3
2250 MT= INT (10 * MT) / 10:MN = INT (10 * MN) / 10
2260 PRINT : PRINT " NUCLEAR MAGNITUDE=   "MN
2270 PRINT " TOTAL MAGNITUDE= ";MT
2280 PRINT
4990 PRINT "DO YOU WANT ANOTHER DATE? (Y/N): ": C$=INPUT$(1)
4992 IF C$="Y" THEN 130
4993 PRINT " SIGNING OFF"
4999 END
5000 REM **CALCULATES JULIAN DAY**
5010 IF MM<= 2 THEN X = 1
5020 IF MM > 2 THEN X= 0
5030 IF S=0 THEN C=2
```

```
5040 IF S=1 THEN C= INT ((YY - X) / 100)
5050 A1= INT ( INT (365.25 * (YY - X)) - .75 * C)
5060 B = INT (367 * ((MM - 2) / 12 + X))
5070 IF B < 1 THEN B = B + 1
5080 JD = 1721088.5# + DD + A1 + B
5090 RETURN
6000 REM
6005 REM*** SETS UP VECTORS TO CONVERT FROM ORBITAL ***
6010 REM ***         TO ECLIPTIC COORDINATES         ***
6015 REM
6020 PX = COS (O2) * COS (O1)- SIN (O2) * SIN (O1) * COS (IN)
6030 PY = COS (O2) * SIN (O1) + SIN (O2) * COS (O1) * COS (IN)
6040 PZ = SIN (O2) * SIN (IN)
6050 QX= - SIN (O2) * COS (O1)- COS (O2) * SIN (O1) * COS (IN)
6060 QY = - SIN (O2) * SIN (O1) + COS (O2) * COS (O1) * COS (IN)
6070 QZ = COS (O2) * SIN (IN)
6090 RETURN
7000 REM *************************
7010 REM   ITERATIVE SOLUTION OF
7020 REM      KEPLER'S EQUATION
7030 REM *************************
7060 X = M/TwoPi
7070 M  = (X- INT (X)) * TwoPi
7080 EA = M
7090 EB=EA-(EA-M-E * SIN (EA)) / (1#-E * COS (EA))
7100 IF ABS(EB-EA)/EB < (.0000001#) THEN 7130
7110 EA=EB
7120 GOTO 7090
7130 RETURN
7500 REM POSITION OF THE SUN
7510 REM G1=MEAN ANOMALY; OP=PERIHELION LONG; L1= SUN'S LONGITUDE
7520 D1 = (JD-2415020&) + JH
7530 X1= (358.4758 + .9856 * D1) / 360
7550 G1 = (X1- INT (X1)) * 360 / RD
7560 OP = (281.2208 + .000047 * D1) / RD
7570 L1=G1 + .03344 * SIN (G1) + .000349 * SIN (2 * G1) + OP
7580 X1 = L1 / TwoPi
7585 REM CALCULATE FINAL L1 AND PRECESS
7586 REM THEN SUN AND COMET ARE REFERRED TO SAME EPOCH
7590 L1 = (X1 - INT (X1)) * TwoPi + .0002437 * (2433282.5# - JD) / 365.25
7595 REM SE=SUN—EARTH DISTANCE;XS,YS,ZS=RECT. COORD. OF SUN
7600 SE=.9997200000000001# / (1# + .01675 * COS (L1-OP))
7610 XS = SE * COS (L1)
7620 YS = SE * SIN (L1) * COS (OB)
7630 ZS = SE * SIN (L1) * SIN (OB)
7640 REM   ---     RA=RIGHT RSCENSION OF SUN——
7645 REM   ---     DS=DECLINATION OF SUN——
7650 SA = ZS / SE:CA= SQR (1#-SA * SA): GOSUB 8500
7660 DS=A3:SA = YS / (SE * CA):CA = XS / (SE * CA): GOSUB 8500
7670 AS1=A3
7800 RETURN
8000 REM CALCULATES COORDINATES IN
8010 REM ECLIPTIC SYSTEM
8020 X1 = A * (COS (EA)-E)
8030 Y1 = A * SQR (1# - E * E) * SIN (EA)
8040 X = X1 * PX + Y1 * QX
8050 Y = X1 * PY + Y1 * QY
8060 Z = X1 * PZ + Y1 * QZ
```

```
8065 REM ***ROTATE TO EQUATORIAL SYSTEM
8070 XC = X:YC = Y * COS (OB)-Z * SIN (OB):ZC = Y * SIN (OB) + Z * COS(OB)
8080 RETURN
8500 REM
8505 REM TWO ARGUMENT ARCTAN
8510 REM CA=COS(A);SA=SIN(A); RETURNS A IN PROPER QUADRANT
8515 REM
8520 IF ABS (CA) = (.00000001#) THEN 8540
8530 A3 = ATN (SA / CA): GOTO 8560
8540 IF SA > 0 THEN A3 = .5# * Pi
8550 IF SA < 0 THEN A3 = 1.5# * Pi
8560 IF CA < 0 THEN 8590
8570 IF CA > 0 AND SA < O THEN 8600
8580 GOTO 8610
8590 A3 = A3 + Pi: GOTO 8610
8600 A3 = A3 + 2# * Pi
8610 RETURN
9000 REM CALCULATE SIDEREAL TIME
9010 T1 = (JD-2415020#) / 36525#: REM CENTURIES SINCE 1900
9020 X1=(18.64606# + 2400.0513# * T1) / 24 + .5 + JH
9030 ST = (X1- INT (X1)) * TwoPi: REM SIDEREAL TIME IN RADIANS
9040 RETURN
```

```
------POSITION OF COMET------                ------POSITION OF COMET------

 Date: 2/ 1/1986  Julian Day: 2446462.000     Date: 1/ 1/1992  Julian Day: 2448622.000
 Local Time: Solar = 0: 0  Sidereal = 8:39    Local Time: Solar = 0: 0  Sidereal = 6:35

   COORDINATES OF THE SUN:                      COORDINATES OF THE SUN:
     X = 0.6506 Y = -0.6789 Z = -0.2944          X = 0.1603 Y = -0.8901 Z = -0.3859
     R.A. = 20 55 DEC = -17 22                   R.A. = 18 40 DEC = -23 6

   COORDINATES OF THE OBJECT:                   COORDINATES OF THE OBJECT:
     X = 0.5162 Y = -0.3303 Z = 0.0612           X = -11.9644 Y = 10.2852 Z = -0.7706
     R.A. = 21 16 DEC = -8 35                    R.A. = 9 26 DEC = -4 24
     R = 0.62 Delta = 1.56                       R = 15.80 Delta = 15.13
     Elongation of Body = 10.2                   Elongation of Body = 131.2
         Hour Angle = 11 23 West                     Hour Angle = 21 9 West
         Azimuth = 342 49                            Azimuth = 128 8
         Elevation = -57 31                          Elevation = 30 49

 Do you want another calculation? (Y/N):      Do you want another calculation? (Y/N):
```

Figure B.3 Two sample results from the Pascal program, using the orbital elements input on page 234 (bottom) for Halley's comet.

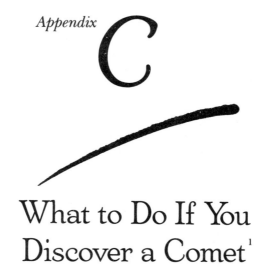

Appendix

C

What to Do If You
Discover a Comet[1]

A better but longer title for this article would be "Notes on Proper Follow-up Procedures If You *Think* You've Discovered a Comet." Why? Because, based on reports received at the Central Bureau for Astronomical Telegrams (CBAT), the word "discover" has different meanings depending on whether it's used by an experienced comet observer or by one who's just beginning. And this difference is visible in numbers: for every real comet discovery, there are perhaps five that do not pan out. In most cases the erroneous claims come from observers who are neither comet hunters nor experienced observers. They are frequently backed by just a single sighting (sometimes a suspicious-looking smudge on a photograph) with no sign of motion.

If you think you've found a comet, here's a checklist to follow before proceeding further:

1. Reprinted from the October 1987 issue of *Sky & Telescope,* pages 420–421. Courtesy of *Sky & Telescope* and Daniel W. E. Green, Smithsonian Astrophysical Observatory.

• Be absolutely sure the image is real. A common problem occurs when visual and photographic observers discover a ghost image—a bit of spurious nebulosity caused by the scattered light from a bright star, planet, or other object in, or just outside, the instrument's field of view. Even very experienced professional astronomers have been fooled. To rule out a ghost, change your eyepiece and move the telescope slightly so that the suspect moves from one edge of the field to the other. If the object does not stay fixed relative to the stars, it is not a comet. Photographers should never rely on a single image—more than two exposures on more than one night are preferred. A trailed one on a long exposure is more promising, but still not proof. Also, the report should be made within one to two days of the last confirmation photograph; possible discoveries that are reported weeks or months later are usually impossible to confirm.

• Use a high power to see if your suspect resolves into a faint grouping of stars; at low powers such an asterism often looks fuzzy.

• Check a good atlas for a galaxy, nebula, or cluster in the suspect's position. Many false alarms come from observers who fail to realize that their "comet" is an entity far beyond our solar system. There are thousands of fuzzy-looking objects within reach of ordinary amateur instruments. Modern charts like those in *Sky Atlas 2000.0* plot many deep-sky objects that can be confused with comets, but even that atlas has omissions. Refer to detailed catalogues such as the *New General Catalogue*, the *Uppsala General Catalogue,* or the Palomar Observatory *Sky Survey* for objects fainter than 11th magnitude.

• Is there motion? If you check several atlases and still find no known object, return to the telescope and carefully sketch the field. Measure the suspect's position to the nearest arc minute in declination and 0.1 minute of time in right ascension if possible (1950 coordinates are preferred). Such accuracy helps others trying to confirm your discovery (poor accuracy can even cause the object to be lost). Be sure to note the date and time of each observation (use Universal, *not* civil, time).

Wait an hour or more and sketch the field again. If the object has moved, record its direction and the precise distance it traveled in the elapsed time interval. Be skeptical if your.find stays put.

• Estimate the total brightness. For a diffuse object, compare the in-focus image with out-of-focus stars until you find some that match it in size and intensity. For bright comets, the *AAVSO Variable Star Atlas* is a good source of accurate stellar magnitudes. Also note the object's angular size, diffuseness, and degree of central condensation. If you see a tail, be sure to record its length and orientation.

A separate sighting on a second night is always recommended *before* you report the object. This has been a standard policy of the prolific comet discoverer William Bradfield of Australia. It is a good idea *first* to ask an experienced amateur comet observer for confirmation before reporting it further.

Use the time between observations to check the published lists of known comets. On any given night there are usually two or more visible in an 8-inch telescope. Good sources of information are the International Astronomical Union (IAU) *Circulars,* the annual *International Comet Quarterly Comet Handbook,* the annual *Handbook* of the British Astronomical Association, and *Sky & Telescope*'s Comet Digest.

The CBAT also has a computer service accessible by modem through telephone lines, Telenet, or SPAN, whereby messages can be left, search ephemerides can be computed (to see if a possible comet is already known), and the IAU *Circulars* can be read.

The CBAT also telegraphs urgent discovery announcements dispatched to a small group of observers worldwide. The more detailed postcard-size *Circulars* themselves are airmailed to a much larger list of subscribers. Write to the CBAT for further information about these services.

How to Send a Report

If your object passes all the various tests, you should send a telegram to the CBAT's number: TWX 710-320-6842 (answerback ASTROGRAM CAM). See the box on pages 254 and 255 for an example. Be sure to include your full name, street address, and telephone number where you can be reached. Give your suspect's right ascension and declination, the dates and times of observations, magnitude, and appearance. Also men-tion your observing location, type of telescope, aperture, and magnification. In the case of photographs, state the film type and exposure times. Subscribers can also send messages to the CBAT computer service already mentioned.

As a backup, be sure to airmail a detailed written account to the postal address given below.

DANIEL W. E. GREEN
Central Bureau for
Astronomical Telegrams
Smithsonian Astrophysical Observatory
60 Garden St.
Cambridge, Mass. 02138

A Sample Telegram

The telegram at right illustrates the preferred way to report a comet discovery to the Central Bureau. It is an example of the official IAU code, long used for astronomical discoveries of many different types.

In this purely hypothetical case, George Smith is reporting a comet discovered in the constellation Sextans by Elmer Jones. The sighting occurred on March 23rd of this year. Jones was able to glimpse a short tail, but he saw no sign of a sharply defined nucleus. The telegram is shown just as we might receive it at the Central Bureau.

The first line gives the discoverer's name (Jones), type of object, and the names of the observers (in this case, both Jones and Smith). Then come two observations of the comet, each represented by a line of five-digit numbers. The first line is deciphered as follows:

19501 gives the equinox of the comet's coordinates, together with a digit indicating the accuracy. Smith is about to report 1950 positions, and the final 1 means they are only rough, probably scaled from a star atlas. The digit 2 is used for a precise position.

70323 tells the date of the observation in year-month-day format. Only the final digit of the year is stated. Since March is the third month (03) of the year, we read the date as "1987 March 23rd."

25000 is the time of the observation, given to five decimals of a 24-hour day. This observation was made at $6^h 00^m$ Universal time, and Smith divided 6 by 24 to get 25000.

09589 is the comet's right ascension, in hours (first two digits) and minutes to the nearest 0.1 minute, omitting the decimal point. Here, then, right ascension $9^h 58^m.9$ is reported.

20022 gives the declination expressed in degrees and minutes of arc. The initial digit is always either 1 or 2, where 1 means − and 2 means +. Therefore, 20022 is $+ 0°22'$.

01115 is a report on how Comet Jones looked. The first two digits tell the type of magnitude being reported (01 = total, 02 = nucleus only). The next two are the magnitude itself rounded to the nearest whole magnitude, 11 in this example. The last is a code summarizing the comet's appearance, where 0 would mean stellar. Digits 1 through 9 in this location have the meanings summarized in the table on page 255:

```
JONES COMET JONES SMITH
19501 70323 25000 09589 20022 01115 45550 30726
19501 70324 12500 10018 10201 01115 23659 21334
VISUAL OBSERVATIONS WITH 20CM REFLECTOR AT GALAXY MOUNTAIN
NEAR COMET WEATHERBEE
DISCOVERER ELMER JONES
GEORGE SMITH, 23 SKYVIEW LANE, ANYTOWN, FUNNYSIDE ISLAND
PHONE 111-555-1234
```

| | *No* | *Tail* | *Tail* |
Head (coma)	*tail*	*under 1°*	*over 1°*
No report	1	2	3
Diffuse without condensation	4	5	6
Diffuse with condensation	7	8	9

Thus Comet Jones has a tail shorter than 1° and a coma that appears diffuse without a condensed nuclear region.

The seventh and eighth groups are not actual data, but merely checks so that those who receive the telegram can tell if any of the preceding digits have been garbled in transmission: **45550** are the last five figures of the sum of all six preceding groups, and **30726** is a separate check-sum of groups 4, 5, and 6 alone. At the Central Bureau we add up the groups again.

The sample telegram also has a second line of numbers, giving the results of a confirming observation the next night. Then Smith describes the telescope and observing site, adding that the suspect comet is near previously known Comet Weatherbee (which, like all other data in this example, is fictitious). Such comments are extremely helpful to us at the Central Bureau, for they show the observers are aware of other known comets.

Smith then signs his full name and includes a complete postal address and telephone number.

Readers interested in a full list of examples, showing how to report a newly discovered nova, supernova, variable star, or minor planet, as well as a comet, may write to the Central Bureau at the address given in the text above.

DANIEL W. E. GREEN

Appendix D

In Situ Measurements and
Observations of Comets
Giacobini-Zinner and Halley

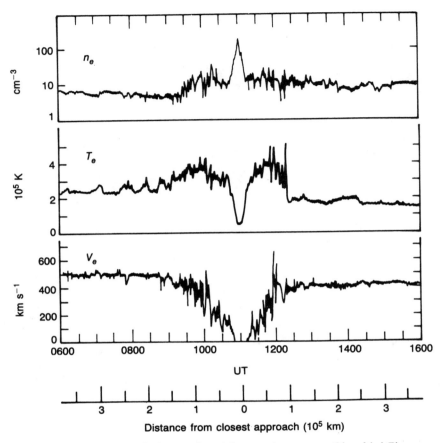

Figure D.1 Overview of plasma data (electrons) at comet Giacobini-Zinner. The general trend is for the plasma to be dense (n_e), cold (T_e), and slowly moving (v_e) at the time of the comet's closest approach to the sun. (Courtesy of S. Bame, Los Alamos National Laboratory)

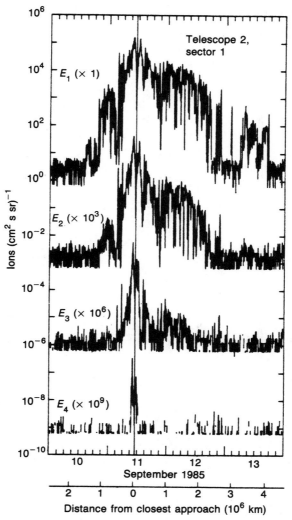

Figure D.2 Energetic ions observed at comet Giacobini-Zinner. The approximate midpoints of the various energy channels in thousands of electron volts (keV) are: $E_1 = 80$; $E_2 = 120$; $E_3 = 170$; $E_4 = 255$; and $E_5 = 395$. The ions are water and are mostly produced by the "pickup" process in the solar wind. Note that the fluxes in the various channels have been scaled. (Courtesy of R. J. Hynds, Blackett Laboratory, Imperial College)

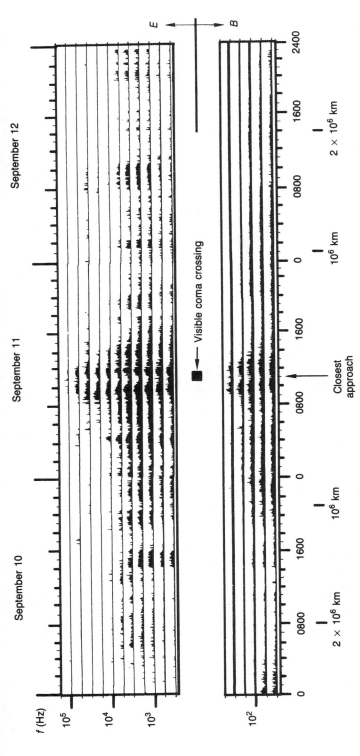

Figure D.3 Overview of plasma wave measurements at comet Giacobini-Zinner. The electric field (E) measurements (top), the peak amplitude, are given in 11 frequency channels and a logarithmic scale. The magnetic field (B) measurements (bottom), the peak amplitude, are given in five frequency channels and a logarithmic scale. The measurements indicate a vast region where the pickup process generates plasma waves. (Courtesy of F. L. Scarf, TRW Space and Technology Group)

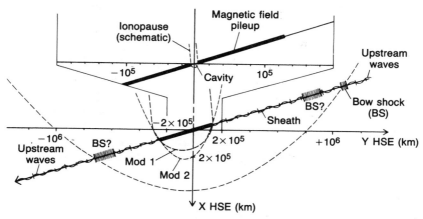

Figure D.4 An overview of the *Giotto* magnetic field measurements at comet Halley. (Courtesy of F. M. Neubauer, Institut für Geophysik and Meteorologie der Universität zu Köln; reprinted by permission from *Nature,* vol. 321, pp. 352–355. Copyright © 1986 Macmillan Magazines Limited)

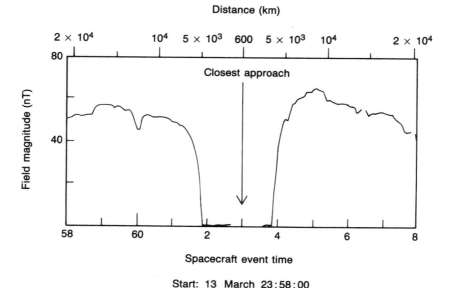

Start: 13 March 23:58:00
End: 14 March 00:08:00

Figure D.5 The field-free cavity as measured at comet Halley. The magnetic field is given in nanoteslas (nT), the same as $\gamma = 10^{-5}$ gauss. (Courtesy of F. M. Neubauer, Institut für Geophysik and Meteorologie der Universität zu Köln; reprinted by permission from *Nature,* vol. 321, pp. 352–355. Copyright © 1986 Macmillan Magazines Limited)

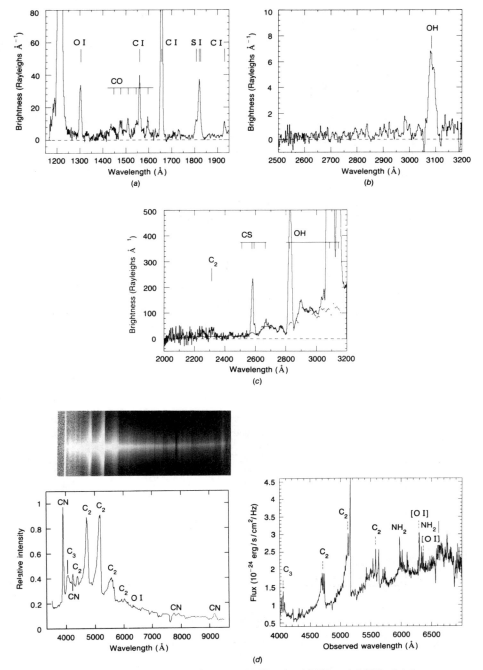

Figure D.6 Sample spectra of comet Halley in 1985 and 1986. *(a) International Ultraviolet Explorer (IUE)* spectrum on March 9, 1986; the very strong feature at 1200 angstroms is the Lyman-alpha line of neutral hydrogen. *(b) IUE* spectrum on September 12, 1985. *(c) IUE* spectrum on March 11, 1986. (*IUE* spectra courtesy of P. D. Feldman, Johns Hopkins University) *(d) Left:* Spectrum of

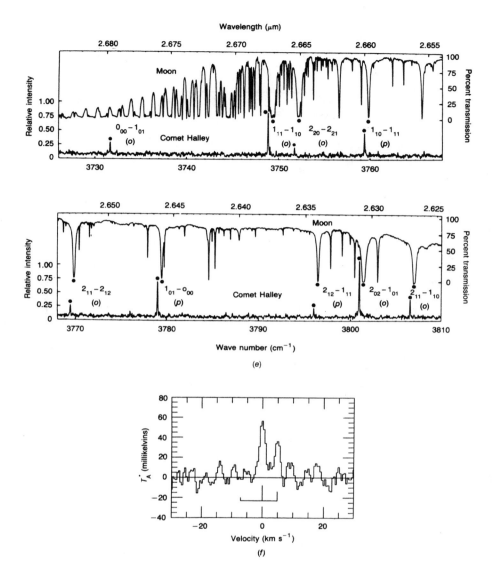

comet Halley on January 6, 1986. (Courtesy of S. M. Larson, Lunar and Planetary Laboratory, University of Arizona) *Right:* Spectrum of comet Halley on March 13, 1986. (Courtesy of H. Spinrad, University of California, Berkeley) *(e)* Infrared spectrum on December 24, 1985, showing lines from ortho *(o)* and para *(p)* water. (Courtesy of M. J. Mumma, NASA-Goddard Space Flight Center) *(f)* Radio wavelength spectrum of hydrogen cyanide (HCN) emission at 3.4 millimeters (88.6 gigahertz); this is the average for March 12, 16, 17, 18, and 21, 1986. The frequency is plotted in terms of the equivalent velocity shift, and the lines beneath the spectrum indicate the position and relative strength of the hyperfine components. (Courtesy of F. P. Schloerb, W. M. Kinzel, D. A. Swade, and W. M. Irvine, Five College Radio Astronomy Observatory, University of Massachusetts)

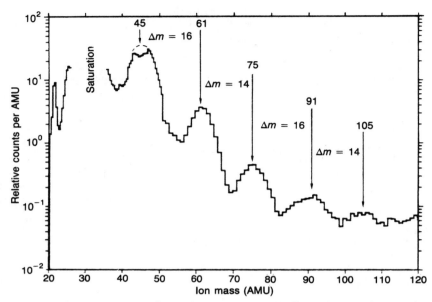

Figure D.7 Mass spectra from *Giotto* at comet Halley. Mass peaks may be attributable to polymerized formaldehyde. (Courtesy of W. F. Huebner, Southwest Research Institute)

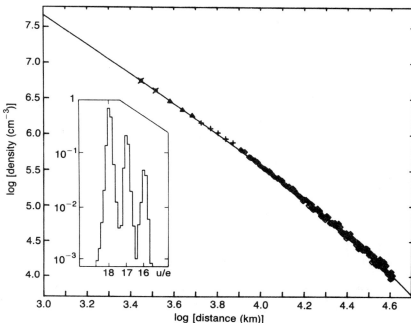

Figure D.8 The variation in the density of water molecules versus distance measured by *Giotto* at comet Halley. Carbon dioxide shows a similar variation near the nucleus. The inset shows the variation for mass per charge 16 to 18, which corresponds to oxygen, hydroxyl, and water, respectively, measured at 2280 kilometers from the nucleus. (Courtesy of D. Krankowsky, Max-Planck-Institut für Kernphysik, Heidelberg)

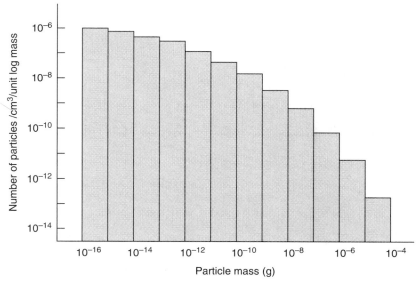

Figure D.9 Mass distribution of dust particles in comet Halley as derived from *Vega-2* in situ measurements. Each bar represents the number of dust particles per cubic centimeter with masses in the range covered by the bar.

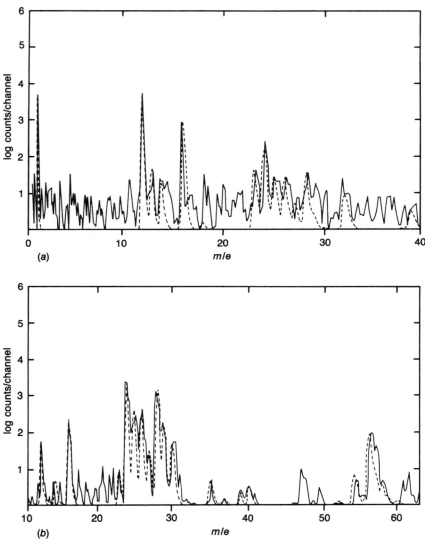

Figure D.10 Sample mass spectra of dust particles in comet Halley as measured by the *Vega* spacecraft: *(a)* A CHON-rich spectrum and *(b)* a silicate-rich spectrum. The solid lines are measured and the dashed lines are fitted under the assumption of normal isotopic composition. In *(a)*, the peaks at *m/e* of 12, 14, and 16 correspond to carbon, nitrogen, and oxygen, respectively. In *(b)*, the peaks at 23, 24, 28, and 56 correspond to sodium, magnesium, silicon, and iron, respectively; these are normal silicate constituents. (Courtesy of J. Kissel, Max-Planck-Institut für Kernphysik, Heidelberg)

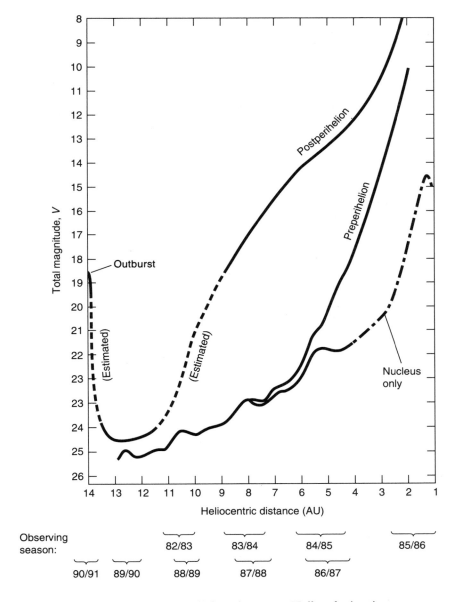

Figure D.11 The brightness variations in comet Halley during its most recent appearance, before and after perihelion. The brightness
is given in terms of the visible magnitude, *V*. The nuclear magnitudes are measurements for the comet's nucleus only. (Adapted
from an illustration by M. J. S. Belton, NOAO)

Glossary

aberration   Here, two similar meanings. First, an effect that results from the fact that light travels at a finite speed. The apparent direction of motion of a photon, seen from a moving object, is the direction of the vector sum of the photon's velocity and the object's velocity; see **Poynting-Robertson effect.** Second, the change in apparent direction of the solar wind caused by a comet's motion.

adsorption   The process by which the atoms or molecules of a gas or liquid adhere to the surface of a solid in a monomolecular layer.

albedo   The fraction of light or electromagnetic radiation that is reflected by a body.

Alfvén speed   The speed at which disturbances move in a magnetized plasma. It is analogous to the sound speed in ordinary gases.

antitail   The taillike structure seen in some comets, pointing toward the sun.

aphelion   The point in the orbit of a celestial body (such as a comet, planet, or asteroid) at which the body is farthest from the sun.

Ariane rocket   A European unmanned liquid-fueled rocket capable of placing scientific payloads in earth orbit.

asteroid   One of the many small bodies orbiting the sun, primarily between Mars and Jupiter; also called **minor planet**.

astronomical unit (AU)   The unit of distance equal to the mean distance between the earth and the sun.

atomic number   A characteristic number of an atom, equal to the number of protons in its nucleus.

atomic weight   A characteristic number of an atom, roughly equal to the number of protons and neutrons in its nucleus.

aurora   A glow from the atoms and ions of the upper atmosphere caused by charged particles from the sun that originate in solar activity; also called **northern lights** (or **southern lights**).

black body   An idealized object that absorbs all photons that strike it; that is, an object that has no reflecting power. A black body has a number of interesting physical properties.

bolide   An especially bright meteor.

bow wave   The shock wave produced when the supersonic solar wind encounters a comet or the magnetosphere of a planet.

Celsius temperature scale (°C)   The temperature scale on which water freezes at 0 degree and boils at 100 degrees at a pressure of 1 atmosphere.

circle   A plane figure traced out by a point that moves such that its distance from a fixed point (its center) is constant.

clathrate hydrate   Water ice with impurity molecules trapped in the crystal lattice.

coma   The cloud of dust and neutral gas surrounding the nucleus of a comet.

cometesimal   A primordial body that may become part of the nucleus of a comet.

conic section   Any of the geometrical figures that are formed by the intersection between a cone and a plane. Circles, ellipses, parabolas, and hyperbolas are conic sections.

contact surface   The surface in the interaction region between a comet and the solar wind separating pure cometary plasma from mixed cometary and solar-wind plasma.

corona   The low-density, hot outer layer of the sun, extending from just above the sun's surface through the solar system.

Cretaceous period   The geological period between 135 million and 65 million years ago; the last period in the Mesozoic (or middle life) era.

Cretaceous-Tertiary boundary (K-T boundary)   The sedimentary layer between the sediments deposited during the Cretaceous and Tertiary geological periods, which marks the precipitous extinction of an enormous number of species, including the dinosaurs.

current sheet   An electrical current in the form of a sheet flowing across the magnetosphere of a planet or the magnetic field of a comet.

Debye length   A characteristic length in a plasma. A microscopic probe in a plasma will detect the charge of a particular ion if it is less than the Debye length away from the ion. If it is farther away, the other ions and electrons will shield the charge of the particular ion.

desorption   The process by which adsorbed material is released from a surface.

disconnection event   Detachment of the tail of a comet from the head. The tail flows away and a new one subsequently forms.

Doppler effect   The change in the pitch of a sound or the color of a light source that occurs when the source and the observer move in relation to each other. A light wave, for example, is shifted toward the red wavelengths if the source and observer are moving away from each other.

eccentricity   A measure of the shape of a conic section. The eccentricity of a circle is zero, and increasing values correspond to ellipses, parabolas (eccentricity of one), and hyperbolas.

ecliptic   The great circle around the sky resulting from the intersection of the sky and the plane of the earth's orbit; the apparent path of the sun around the sky.

ellipse   A plane figure traced out by a point that moves such that the sum of its distances from two fixed points (the foci of the ellipse) is a constant.

flyby mission   A space mission during which a spacecraft flies past a comet, usually at high speed, and makes close observations.

head   The portion of a comet made up of the nucleus and the coma.

high-speed stream   A stream in the solar wind that moves with a speed in excess of the average speed.

hydrogen-hydroxyl cloud   An extensive cloud, composed primarily of atomic hydrogen and the hydroxyl radical, observed to surround many comets.

hydrosphere   The outer liquid portion of the earth, consisting primarily of the oceans.

hydrostatic equilibrium   The condition of a fluid (gaseous or liquid) in which pressure forces counterbalance gravitational forces.

hyperbola   A curve formed by the intersection of a right circular cone and a plane parallel to the axis of the cone.

imaging   The process of forming a picture of an object, so that its shape or changes in its shape can be studied.

inner Oort cloud   A hypothetical, flattened cloud of comets lying between about 1000 and 20,000 astronomical units from the sun.

in situ measurements   Measurements of an object made by probes within the object.

ion   An atom or molecule that has lost or gained one or more electrons and is electrically charged.

ionized gas   A plasma.

isotope   Any of two or more forms of a chemical element with differing numbers of neutrons in the nucleus.

Kelvin temperature scale (K)   The temperature scale on which 0 degree is absolute zero and the difference between the freezing and boiling points of water is 100 degrees. The Kelvin scale and the Celsius scale differ only in their zero points.

Kreutz family   A family of sun-grazing comets, all of which have similar orbital elements and are believed to originate from the breakup of a single, large comet.

Kuiper-Duncan cloud   The innermost, disk-shaped part of the comet cloud, lying between about 35 and 1000 astronomical units from the sun.

lithosphere   The outer rocky shell of the earth, consisting of the crust and the upper mantle.

long-period comet   A comet with a period in excess of 200 years.

magnetosphere The region within the magnetic field of the earth or another planet displaying a variety of plasma physics processes.

meteor The streak of light caused by an interplanetary particle burning up in the earth's atmosphere.

meteorite A solid particle, once a portion of a particle that caused a meteor, that survived to reach the surface of the earth.

meteoroid A solid particle in interplanetary space that can produce a meteor.

neckline structure A narrow, stable feature in the dust tail of some comets.

new start Congressional go-ahead to begin a new space mission (NASA term).

node One of the two points at which the orbit of a comet (or of another body in the solar system) crosses the plane of the ecliptic. The point at which the body crosses the ecliptic from south to north is the *ascending node;* the point at which it crosses the ecliptic from north to south is the *descending node.*

nongravitational force A force resulting from the nonuniform sublimation of ices on the rotating nucleus of a comet. Nongravitational forces can slowly increase or decrease the period of a comet.

nova A star that suddenly appears to increase in brightness by an average factor of 1 million times, then slowly reverts to its original brightness over a few months.

nucleus The solid portion of a comet, believed to be a chunk of dust and ice a few kilometers in diameter.

Oort cloud A cloud of comets surrounding the solar system, roughly between 20,000 and 100,000 astronomical units from the sun.

orbital elements The six parameters that describe the size, shape, and orientation in space of the orbit of a celestial body.

ortho water A water molecule in which the nuclei of the two hydrogen atoms spin in the same direction.

outburst An increase in the brightness of a comet, caused by increased emission of gas and dust.

parabola A figure formed by the intersection of a right circular cone and a plane that is parallel to an element of the cone (a line on the cone passing through its apex); it can be thought of as the limiting case of an ellipse as one focus moves to infinity.

parallax   The apparent shift in the position of a distant object with a change in the position from which it is observed.

para water   A water molecule in which the nuclei of the two hydrogen atoms spin in opposite directions.

parent molecule   A relatively complex molecule that breaks up into simpler, offspring molecules in a comet's atmosphere.

perihelion   The point in the orbit of a celestial body, such as a comet, at which the body is closest to the sun.

photodissociation   The breakup of molecules by sunlight.

photometry   The observational technique by which the brightness of an object is measured.

planetesimal   A primordial body that may become part of a planet.

plasma   A gaseous state of matter composed of positively charged ions and negatively charged electrons of such quantity that the overall plasma is electrically neutral. The behavior of plasmas is influenced by the electrostatic forces between the particles.

plasma instability   Any of the various mechanisms that cause plasma properties to evolve rapidly to new values.

plasma wave   A wave in a magnetized plasma; a phenomenon in which the electric and magnetic properties of a magnetized plasma vary over time in a wavelike manner.

Poynting-Robertson effect   The effect produced by photons that hit dust particles orbiting the sun and causes them to spiral slowly into the sun. See **aberration.**

primitive atmosphere   The original atmosphere of the earth, thought to be composed of simple compounds of hydrogen, such as water, methane, and ammonia.

prograde   The normal direction of motion of the major bodies in the solar system; that is, counterclockwise as seen from far above the earth's north pole.

radiation pressure   (*a*) A pressure caused by photons reflecting off the wall of a container, analogous to the gas pressure produced when the atoms, ions, or electrons of a gas bounce off the wall of a container. (*b*) The force imparted to an object, illuminated on one side only, by the photons reflecting from it.

radical   A piece of a molecule. The OH radical arises from the dissociation of the water molecule ($H_2O = HOH$).

rendezvous mission   A space mission in which a spacecraft flies into the vicinity of a comet and possibly orbits alongside it for a time.

repulsive force   A force exerted by objects on each other, which tends to move them apart, or repel them.

resonance line   A spectrum line produced by a permitted electron transition between the lowest energy level (the ground level) in an atom and the next highest level.

sample return mission   A space mission in which a spacecraft lands on the nucleus of a comet, scoops up a sample of material, and returns with it to earth.

secular acceleration   The small, continuous acceleration of a body such as a comet or the moon along its orbit.

semimajor axis   One-half the longest dimension of an ellipse.

short-period comet   A comet with a period of less than 200 years.

solar electric propulsion   A form of rocket engine in which electric fields generated by solar cells create and accelerate ions to high velocities, to cause a reactive force; also called ion drive.

solar wind   The plasma expanding away from the sun, with a density of about 5 particles per cubic centimeter and a speed of 400 kilometers per second at earth.

spectrograph   An optical device used to study the spectrum emitted by an object.

spectroscopy   The process by which the light from an object is broken up into its constituent wavelengths so that the brightness can be studied as a function of wavelength.

split comet   A comet, such as Biela, whose nucleus is observed to break up into two or more pieces.

sublimation   The process by which a solid material passes into the gaseous state without first becoming a liquid.

sun-grazing comet   A comet whose perihelion point is very close to the sun.

tail   The portion of a comet extending from the head in the direction away from the sun.

tail ray   A narrow, bright feature of a comet extending from the nucleus along the plasma tail.

temperature The measure of the internal energy in a quantity of matter.

Tertiary period The geological period between 65 million and about 2 million years ago; the first period of the Cenozoic (or recent life) era.

tidal forces Forces on a body caused by differing gravitational forces acting on different parts of the body.

Type I tail A plasma tail.

Type II tail A dust tail.

Suggested Readings

Chapter 1: A First Look at Comets

General references:

Brandt, J. C. and Maran, S. P. 1979. *New Horizons in Astronomy*, 2nd ed. New York: W. H. Freeman and Company.

Brandt, J. C. and Chapman, R. D. 1981. *Introduction to Comets*. Cambridge and New York: Cambridge University Press.

Brandt, J. C. 1981. *Comets*. San Francisco: W. H. Freeman and Company. A collection of articles from *Scientific American*.

Chapman, R. D. 1978. *Discovering Astronomy*. New York: W. H. Freeman and Company.

Bortle, J. E. Comet Digest in *Sky & Telescope*. John Bortle is a regular contributer to this magazine aimed at amateur astronomers. He writes about the comets that are currently visible.

Freitag, R. S. 1984. *Halley's Comet: A Bibliography*. Library of Congress. Washington, D.C.: GPO. An extensive bibliography of writing about comet Halley.

Littman, M. and Yeomans, D. K. 1985. *Comet Halley: Once in a Lifetime*. Washington, D.C.: American Chemical Society.

For an introductory astronomy glossary and an up-to-date basic astronomy library, see *The Universe in the Classroom, A Newsletter on Teaching Astronomy*, numbers 14 & 15, Spring 1990, San Francisco: The Astronomical Society of the Pacific.

The following references talk about the early history of comets:

Aristotle, 1952. Meteorology. *Great Books of the Western World* 8:455. Chicago: Encyclopedia Britannica.

Armitage, A. 1966. *Edmond Halley*. London: Thomas, Nelson.

Bede. 1968. *A History of the English Church and People*. New York: Penguin Books.

Pannekoek, A. 1961. *A History of Astronomy*. New York: Interscience.

Yeomans, D. K. 1991. *Comets, A Chronological History of Observation, Science, Myth, and Folklore*. New York: Wiley Science Editions.

For more background on the physics or chemistry discussed in this book, see any up-to-date textbook; for instance:

Sears, F. W., Zemansky, M. W., and Young, H. D. 1980. *College Physics*. Reading, Mass.: Addison-Wesley Publishing Co.

Timberlake, K. 1988. *Chemistry*. New York: Harper and Row.

Chapter 2: The Orbits and Motions of Comets

More detail on the later history of comets is discussed in the following:

Hellman, C. D. 1944. *The Comet of 1577: Its Place in the History of Astronomy*. New York: AMS Press.

Pannekoek, A. 1961. *A History of Astronomy*. New York: Interscience.

Information on the orbits and dynamics of comets can be found in the following:

Marsden, B. G. 1968. Comets and nongravitational Forces, I. *Astron. J.* 73:367.

Marsden, B. G. 1968. Comets and nongravitational Forces, II. *Astron. J.* 74:720.

Marsden, B. G. 1968. Comets and nongravitational Forces, III. *Astron. J.* 75:75.

Marsden, B. G. 1989. *Catalogue of Cometary Orbits,* 6th ed. Cambridge, Mass.: S.A.O. Minor Planet Center.

Nakano, S. and Green, D. W. E. 1991. *Comet Handbook of the International Comet Quarterly.* Lists ephemerides of comets for 1991.

Other references:

Marsden, B. G. 1989. The sungrazing comet group, II. *Astron. J.* 98:2306.

Chapter 3: The Heads of Comets

The nature of water ice and clathrate hydrates:

Atkins, P. W. 1987. *Physical Chemistry.* New York: W. H. Freeman and Company.

The development of the "dirty snowball" model:

Whipple, F. L. 1950. A comet model, I. The acceleration of comet Encke. *Astrophys. J.* 111:375.

Whipple, F. L. 1950. A comet model, II. Physical relations for comets and meteors. *Astrophys. J.* 113:464.

Whipple, F. L. 1955. A comet model, III. The zodiacal light. *Astrophys. J.* 121:750.

Outbursts:

Donn, B. and Urey, H. C. 1965. On the mechanism of comet outbursts and the chemical composition of comets. *Astrophys. J.*123:339.

Hughes, D. W. 1975. Cometary outbursts: a brief survey. *Quart. J. R. Astron. Soc.* 16:410.

Other relevant references:

Delsemme, A. H. and Swings, P. 1952. Hydrates de gaz dans les noyaux cometaires et les grains interstellaires. *Ann. Astrophys.* 15:1.

Donn, B., Mumma, M., Jackson, W., A'Hearn, M., and Harrington, R. (eds.). 1976. *The Study of Comets.* NASA SP-393. Washington, D.C.: GPO.

Whipple, F. L. 1976. Background of modern comet theory. *Nature* 263:15.

Wilkening, L. L. (ed.). 1982. *Comets*. Tucson: The University of Arizona Press. There are numerous relevant articles in this volume, with extensive bibliographies.

Yeomans, D. K. and Chodas, P. W. 1989. An asymmetric outgassing model for cometary nongravitational accelerations. *Astron. J.* 98:1083.

Also see the summary/conference volumes given in the readings for Chapter 6.

Chapter 4: The Tails of Comets

Bittencourt, J. A. 1986. *Fundamentals of Plasma Physics*. New York: Pergamon Press. The basic plasma physics relevant to cometary science is presented in this reference.

Relevant references:

Brandt, J. C. 1968. The physics of comet tails. *Ann. Rev. Astron. Astrophys.* 6:267

Donn, B., Mumma, M., Jackson, W., A'Hearn, M., and Harrington, R. (eds.). 1976. *The Study of Comets*. NASA SP-393. Washington, D.C.: GPO.

Wilkening, L. L. (ed.). 1982. *Comets*. Tucson: The University of Arizona Press. There are numerous relevant articles in this volume, with extensive bibliographies.

Also see the summary/conference volumes given in the readings for Chapter 6.

Chapter 5: Observing Comets

Arpigny, C. 1965. Spectra of comets and their interpretation. *Ann. Rev. Astron. Astrophys.* 3:351.

Green, D. W. E. (ed.). *International Comet Quarterly*. Cambridge, Mass.: The Smithsonian Astrophysical Observatory. This periodical is devoted to news and observations of past and present comets.

Swings, P. 1943. Cometary spectra. *Mon. Not. R. Astron. Soc.* 103:86.

Wilkening, L. L. (ed.). 1982. *Comets*. Tucson: The University of Arizona Press. There are numerous relevant articles in this volume, with extensive bibliographies.

Also see the summary/conference volumes given in the readings for Chapter 6.

Chapter 6: In Pursuit of Comets Giacobini-Zinner and Halley

20th ESLAB Symposium on the Exploration of Halley's Comet. Vol. 1, *Plasma & gas.* Vol. 2, *Dust & nucleus.* Vol. 3, *Posters & late papers.* ESA SP-250. Noordwijk: ESA Publication Division.

Alfvén, H. 1957. On the theory of comet tails. *Tellus* 9:92.

Birmingham, T. J. and Dessler, A. J. (eds.). 1988. *Comet Encounters,* Washington, D. C.: American Geophysical Union.

Brandt, J. C. 1990. The large-scale plasma structure of Halley's comet, 1985–1986. In *Comet Halley,* Vol. 1. Mason, J. (ed.). New York: Ellis Horwood, pp. 33–55.

Donn, B., Rahe, J., and Brandt, J. C. 1988. *Atlas of Comet Halley 1910 II.* NASA SP-488. Washington, D.C.: GPO.

Encounters with comet Halley: the first results. *Nature* 321 (15 May 1986). A collection of papers on the first results from the spacecraft that flew by Halley.

Farquhar, R. W. 1988. Teaching old spacecraft new tricks. *Sky & Telescope* 76:134.

Farquhar, R. W., Muhonen, D., and Church, L. C. 1985. Trajectories and orbital maneuvers for the ISEE-3/ICE comet mission. *J. Astronaut. Sci.* 33:235.

Grewing, M., Praderie, F., and Reinhard, R. (eds.). 1988. *Exploration of Halley's Comet.* Berlin: Springer-Verlag. This is a collection of papers on the scientific results from Halley's comet.

Mason, J. (ed.). 1990. *Comet Halley, Investigations, Results, Interpretations.* Vol. 1, *Organization, plasma, gas.* Vol. 2, *Dust, nucleus, evolution.* New York: Ellis Horwood.

Mendis, D. A. 1988. A postencounter view of comets. *Ann. Rev. Astron. Astrophys.* 26:11.

Newburn, R. L., Jr., Neugebauer, M., and Rahe, J. (eds.). 1991. *Comets in the Post-Halley Era.* Vol. 1 and Vol. 2. Dordrecht: Kluwer Academic Publishers.

Reinhard, R. and Battrick, B. (eds.). 1986. *Space Missions to Halley's Comet.* ESA SP-1066. Noordwijk: ESA Publications Division.

von Rosenvinge, T. T., Brandt, J. C., and Farquhar, R. W. 1986. The International Cometary Explorer mission to comet Giacobini-Zinner. *Science* 232:353.

Chapter 7: The Origins of Comets

Bailey, M. E., Clube, S. V. M., and Napier, W. M. 1990. *The Origin of Comets*. Oxford: Pergamon Press.

Delsemme, A. H. (ed.). 1977. *Comets, Asteroids, Meteorites: Interrelations, Evolution, and Origins*. Toledo: University of Toledo.

Duncan, M., Quinn, T., and Tremaine, S. 1987. The formation and extent of the solar system comet cloud. *Astron. J.* 94:1330.

Duncan, M., Quinn, T., and Tremaine, S. 1987. Origins of short period comets. *Astrophys. J.* 328:L69.

Oort, J. H. 1950. The structure of the cloud of comets surrounding the solar system and a hypothesis concerning its origin. *Bull. Astron. Inst. Neth.* 11:91.

Chapter 8: Comets and the Solar System

Alvarez, L. W., Alvarez, W., Asaro, F., and Michel, H. V. 1980. Extraterrestrial cause for the Cretaceous-Tertiary extinction. *Science* 208:1095–1108.

Biermann, L. 1951. Kometenschweife und Korpuskularstrahlung. *Z. Astrophys.* 29:274.

Biermann, L. 1953. Physical processes in comet tails and their relation to solar activity. *Mem. Soc. Sci. Liège, 4th Ser.* 13:251.

Dessler, A. J. 1991. The small-comet hypothesis. *Reviews of Geophysics* 29:355–382; 29:609–610.

Frank, L. A. 1990. *The Big Splash*. Secaucus, N.J.: Carol Publishing Group.

Gehrles, T. 1979. *Asteroids*. Tucson: University of Arizona Press.

Lagerkvist, C.-I. and Rickman, H. (eds.). 1984. *Asteroids, Comets, Meteors*. Uppsala: University of Uppsala.

Lagerkvist, C.-I., Lindblad, B. A., Lundstedt, H., and Rickman, H. (eds.). 1986. *Asteroids, Comets, Meteors II*. Uppsala: University of Uppsala.

Ponnamperuma, C. (ed.). 1981. *Comets and the Origin of Life*. Dordrecht: D. Reidel.

Sekanina, Z. 1983. The Tunguska event: no cometary signature in evidence. *Astron. J.* 88:1382–1414.

Shoemaker, E. M. and Wolfe, R. F. 1986. Mass extinctions, crater ages and comet showers. In *The Galaxy and the Solar System*. Smoluchowski, R., Bahcall, J. N., and Matthews, M. S. (eds.). Tucson: University of Arizona Press, pp. 338–386.

Chapter 9: The Future

Comet Nucleus Sample Return. ESA Special Publications. Noordwijk: ESA Publications Division.

Farquhar, R. W., Dunham, D. W., and Hsu, S. C. 1987. A *Voyager*-style tour of comets and asteroids. *J. Astronaut. Sci.* 35:399.

French, L. M., Vilas, F., Hartmann, W. K., and Tholen, D. J. 1990. Distant asteroids and Chiron. In *Asteroids II*. Binzel, R. P., Gehrles, T., and Matthews, M. S. (eds). Tucson: University of Arizona Press, pp. 468–486.

Kuiper, G. P. and Roemer, E. 1972. *Comets: Scientific Data and Missions*. Tucson: University of Arizona, Lunar and Planetary Laboratory.

McGlynn, T. A. and Chapman, R. D. 1989. On the non-detection of extrasolar comets. *Astrophys. J.* 346:L105.

Meech, K. J. and Belton, M. J. S. 1990. The atmosphere of 2060 Chiron. *Astron. J.* 100:1323.

Neugebauer, M., Yeomans, D. K., Brandt, J. C., and Hobbs, R. W. (eds.). 1979. *Symposium on Space Missions to Comets*. NASA Conference Publication 2089. Washington, DC: GPO.

Neugebauer, M. and Weismann, P. R. 1989. The CRAF mission. *EOS* 70:633.

Solar System Exploration Committee. 1988. *Planetary Exploration Through the Year 2000: Scientific Rationale*. Washington, D.C.: GPO.

Stern, S. A. 1989. Implications of volatile release from object 2060 Chiron. *Publications of the Astronomical Society of the Pacific* 101:126.

Stern, S. A., Shull, J. M., and Brandt, J. C. 1990. Evolution and detectability of comet clouds during post-main-sequence stellar evolution. *Nature* 345:305.

Appendix A: Useful Equations

Combi, M. R. and Delsemme, A. H. 1980. Neutral cometary atmospheres, I. An average random walk model for photodissociation in comets. *Astrophys. J.* 237:663.

Combi, M. R. and Delsemme, A. H. 1980. Neutral cometary atmospheres, II. The production of CN in comets. *Astrophys. J.* 237:641.

Festou, M. M. 1981. The density distribution of neutral compounds in cometary atmospheres. *Astron. Astrophys.* 95:69.

Harwit, M. 1988. *Astrophysical Concepts*. Berlin: Springer-Verlag.

Haser, L. 1957. Distribution d'intensite dans les tete d'une comete. *Bull. Acad. R. Belg., 5e serie.* 43:740.

Lang, K. R. 1980. *Astrophysical Formulae*. Berlin: Springer-Verlag.

Appendix B: How to Calculate Positions of a Comet

Brouwer, D. and Clemence, G. M. 1961. *Methods of Celestial Mechanics*. New York: Academic Press.

Meeus, J. (1988). *Astronomical Formulae for Calculators*. Richmond, Va.: Willmann-Bell.

Tattersfield, D. 1984. *Orbits for Amateurs with a Microcomputer*. Cheltenham: Stanley Thornes (Publishers) Ltd. This book takes you step-by-step through the complexities of orbit calculations and provides BASIC programs for each step.

Appendix D: In Situ Measurements and Observations of Comets Giacobini-Zinner and Halley

See references for Chapter 6.

Index

Comet Index